Shock Waves in Solid State Physics

Shock Waves in Solid State Physics

G. I. Kanel'

CISP

CRC Press
Taylor & Francis Group
Boca Raton London New York

CRC Press is an imprint of the
Taylor & Francis Group, an **informa** business

Translated from Russian by V.E. Riecansky

CRC Press
Taylor & Francis Group
6000 Broken Sound Parkway NW, Suite 300
Boca Raton, FL 33487-2742

First issued in paperback 2021

ISBN 13: 978-1-03-224011-4 (pbk)
ISBN 13: 978-0-367-22589-6 (hbk)

Publisher's Note
The publisher has gone to great lengths to ensure the quality of this reprint but points out that some imperfections in the original copies may be apparent.

Visit the Taylor & Francis Web site at
http://www.taylorandfrancis.com

and the CRC Press Web site at
http://www.crcpress.com

Contents

Introduction

Experiments with the compression of a substance in strong shock waves were started in the forties and fifties of the last century, when information about the equations of state of active and structural materials in the megabar (millions of atmospheres) pressure range was needed to design an atomic bomb. The equations of state still remain one of the main problems of the physics of shock waves in condensed matter, but with the development of experimental techniques and the accumulation of the information obtained, new, equally important and interesting directions have emerged, which are not always related to defense topics. Currently, researchers have at their disposal a well-developed set of methods for generating, diagnosing, and interpreting shock-wave phenomena in condensed media, using the experimental information obtained on the elastoplastic and strength properties of industrial metals and alloys, geological materials, ceramics, glasses, polymers and elastomers, ductile and brittle single crystals in the microsecond and nanosecond ranges of shock load durations; on the parameters and kinetic laws of the physicochemical transformations of a substance during its shock compression; on the properties of various substances in states far from equilibrium, which are realized at extremely short durations of shock-wave action.

The wave nature of the load makes the interpretation of the measurement results quite visual and unambiguous. The measurements are based on the fact that the structure of the waves and the dynamics of wave interactions are determined by the processes of elastoplastic deformation, physicochemical transformations and failure in the material. These processes are accompanied by changes in the compressibility of the material, which in turn leads to the formation of specific features in the structure of intense compression and rarefaction waves. Information on the properties of the tested materials is found from the measurement results by both direct processing of the obtained wave profiles and their comparison with the results of mathematical modelling of shock wave phenomena.

In the latter case, the properties of the medium are described by the constitutive relations, which are, as a rule, semi-empirical in nature and generalize the experimental data on the basis of certain theoretical ideas about the behaviour of materials. The data obtained in this way on the properties of model and structural materials are then used in calculations of the functioning of various technical devices under the conditions of intense impulse effects, as well as for the development of physical theories of strength and plasticity. However, it should be noted that, despite a completely satisfactory general understanding of the physics and mechanics of high-speed deformation and fracture, a detailed agreement of theoretical concepts and models of these phenomena with the available experimental data has not yet been achieved.

The main purpose of the investigation of shock-wave phenomena in solids is to ensure the predictability of the action of an explosion, high-speed impact, laser and other intense pulsed effects on materials and structures. In the modern sense, comprehensive predictability is achieved by computer simulation of the processes under consideration, for which we need thermodynamic equations of state describing the relationship between pressure, density, phase composition and internal energy of a substance, as well as the constitutive relations describing the processes of chemical, phase and polymorphic transformations, elastoplastic strains and failure in terms compatible with the conservation equations and the equation of state.

On the other hand, experiments with shock waves make it possible to obtain a unique information about the most fundamental strength properties of materials. The high rate of application of the load allows one to create larger stresses in the material and thereby activate new deformation and fracture mechanisms. The reflection of the shock wave from the surface of the body causes the fracture under the stress state, close to three-dimensional tension, and in the absence of influence of the body surface and the environment. In this case, the short duration and high amplitudes of the shock load make it possible to achieve high stresses in the material and, thus, to move from the consideration of single cracks to an analysis of the evolution of scattered fractures. With such rapid effects, the resistance to fracture of solids becomes comparable with the ultimate theoretical strength determined directly by the potential of interatomic interactions. Thus, the significance of research into the processes of inelastic deformation and fracture of solids under shock-wave loading is determined both by the unique possibility of research in the field of strength and plasticity at the highest and reliably measurable strain rates, and by various practical needs not limited only by shock effects. In a sense, shock wave tests are like a 'time microscope', giving access to elementary acts of deformation and fracture. In the next decade, we should expect a significant expansion of the use of shock wave techniques for solving problems in solid state physics, materials science, strength and plasticity physics.

The book consists of two unequal parts. The first part outlines the main theoretical principles (Chapter 1) and gives an idea of the shock-wave experiment technique (Chapter 2) – with an emphasis on the physical principles underlying the various diagnostic methods, as well as the theoretical justification of studies of dynamic strength and polymorphism in shock waves (Chapter 3). The author believes that the methodological issues are important to justify the reliability and a better understanding of the research results. The second part (Chapters 4–8) presents a number of recent research results that give an idea of the current state of this field of science. Chapter 8 is devoted to the study of the macrokinetic patterns of the response of solid explosives in shock waves and falls somewhat out of the main direction related to the properties of solids. This material, however, is important to illustrate the capabilities of the method. The citation of original works in the text is minimized in order not to impede the perception of the main material. More detailed bibliography can be found in reviews and monographs [3, 4, 14], as well as in the references at the end (references [78–83]).

The author hopes that this book will be useful to specialists in the fields of physics and mechanics of strength and plasticity, physics of structural transformations under pressure, physics of high-speed impact and explosion, as well as undergraduate and graduate students studying physics and mechanics of explosion and high-speed impact.

1

One-dimensional movements of a compressed medium

This chapter provides brief information about the basic state parameters and the laws of motion of continuous compressible media to the extent necessary to understand the peculiarities of the propagation of shock and rarefaction waves in solids, taking into account their elastic–plastic behaviour and possible polymorphic transformations, as well as for discussion of dynamic experiments. It also analyzes the wave dynamics of spallation phenomena – the fracture of solids and liquids under the action of tensile stresses upon reflection of a compression pulse from the surface. A more complete and consistent presentation of the laws of motion of a compressible substance can be found in well-known monographs [1, 2].

To interpret experiments with shock waves, it is sufficient to consider the one-dimensional motion of matter, since it is in this most simple statement for analysis that most of the measurements are carried out. Since the recording of the kinematic parameters of the shock-wave process in a condensed medium is carried out, as a rule, for selected material sections of the sample, it is convenient to analyze wave processes in the substantial Lagrange coordinates associated with the substance. We will use the spatial coordinate x of the particle at the initial moment of time as the Lagrangian coordinate h

$$h = \left(\int_0^x \rho \, dx \right) / \rho_0, \quad \left. \frac{\partial h}{\partial x} \right|_t = \frac{\rho}{\rho_0}, \tag{1.1}$$

where ρ_0, ρ are the density values of the substance at the initial time t_0 and at time t, respectively. The partial derivatives with respect to time t and coordinate h will be denoted as $\frac{\partial}{\partial t} \equiv \left(\frac{\partial}{\partial t}\right)_h$ and $\frac{\partial}{\partial h} \equiv \left(\frac{\partial}{\partial h}\right)_t$; derivatives of the function f along certain trajectories on the $h - t$ plane are expressed by the relations

$$\frac{df}{dt} = \frac{\partial f}{\partial t} + \frac{\partial f}{\partial h}\frac{dh}{dt}, \quad \frac{df}{dh} = \frac{\partial f}{\partial h} + \frac{\partial f}{\partial t} / \frac{dh}{dt}. \tag{1.2}$$

where dh/dt is the slope of the selected trajectory.

1.1. One-dimensional continuous flow of compressible medium

Neglecting the deviator stress components, thermal conductivity, energy release or absorption and relaxation phenomena, the one-dimensional motion of a continuous compressible medium is described by a set of partial differential equations expressing the fundamental laws of conservation of mass, momentum, and energy, supplemented by the equation of state of matter:

$$\rho_0 \frac{\partial V}{\partial t} - \frac{\partial u_p}{\partial h} = 0, \; \rho_0 \frac{\partial u_p}{\partial t} + \frac{\partial p}{\partial h} = 0, \; \frac{\partial E}{\partial t} = -p\frac{\partial V}{\partial t}, \; E = E(p,V) \tag{1.3}$$

where p is the pressure, u_p is the particle velocity (the velocity of the particles of a substance), $V = 1/\rho$ is the specific volume and E is the specific internal energy. The inference of the continuity equation (the first equation of system (1.3)) is illustrated in Fig. 1.1 *a*.

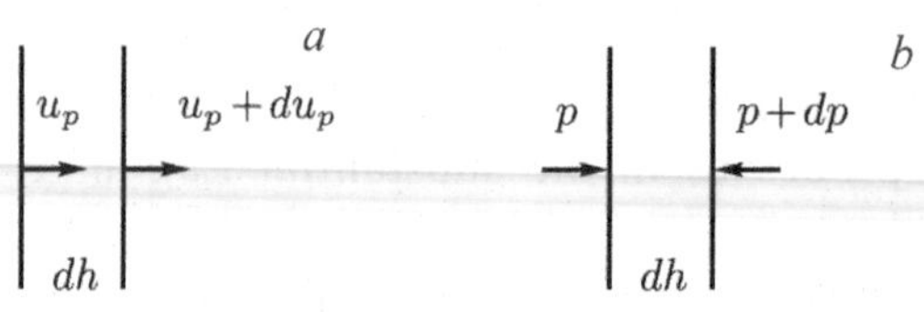

Fig. 1.1. To clarify the inference of the continuity equation and the equation of conservation of momentum.

We consider a layer of substance with a thickness dh with a unit cross-sectional area, the left boundary of which moves with the substance at a speed u_p, and the right one at a speed $u_p + du_p$. The volume of this layer at time t is equal to dh. During the period of time dt, the left boundary of the layer is shifted by a distance u_p

dt, and the right boundary by distance $(u_p + du_p)dt$, as a result of which the volume of the layer becomes equal to $dh + du_p\, dt$. In this case, it is seen from the figure that in the case of a positive velocity gradient ($du_p/dh > 0$), the layer thickness increases with time, and with a negative velocity gradient the selected layer is compressed. The relative change in the volume of the layer dV/V_0 is proportional to the change in the thickness of the selected layer, equal to $du_p\, dt$. Accordingly, the rate of change of volume is $\frac{\partial(V/V_0)}{\partial t} = \frac{\partial u_p}{\partial h}$.

The inference of the equation of conservation of momentum (the second equation of system (1.3)) is explained in Fig. 1.1 *b*. A layer of matter with a thickness dh is again considered with a unit cross-sectional area, on the left border of which the pressure p acts, and on the right one – pressure $p + dp$. The mass of the layer is $\rho_0\, dh$. If there is a pressure difference dp, the layer as a whole speeds up or slows down its movement in accordance with the second Newton law: $\rho_0 dh \frac{du_p}{dt} = -dp$. The minus sign appeared due to the fact that, as can be seen from the figure, in the case of a positive pressure gradient ($dp > 0$) the layer must be decelerated ($du_p/dt < 0$), in the case of a negative pressure gradient the layer is accelerated.

The energy conservation equation (the third equation in set (1.3)) expresses the condition of the process being adiabatic and is the second law of thermodynamics

$$dE = T\,dS - pdV,$$

where S is the specific entropy in the absence of external heat sources. In the presence of heat generation or heat loss, the change in the specific internal energy E of the selected Lagrangian particle occurs both due to the work of compression, which the environment produces on it, and due to the release (or absorption) of energy Q from auxiliary sources:

$$\frac{dE}{dt} = -p\frac{dV}{dT} + \frac{dQ}{dt}.$$

Isentropic processes are accompanied by a change in the temperature of a substance. Isentropic compression of the substance leads to its heating; as the pressure in a substance decreases, its temperature drops. Due to thermal expansion, the specific volume of

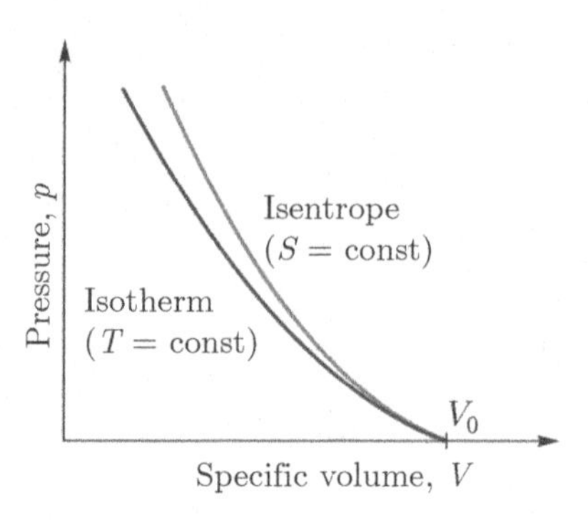

Fig. 1.2. The mutual position of the isentrope and isotherm on the pressure plane, p is the specific volume V.

an adiabatically compressed substance is greater than that compressed at a constant temperature (Fig. 1.2).

The pressure derivative with respect to density along the isentrope is equal to the square of the speed of sound:

$$\left(\frac{dp}{d\rho}\right) = -V^2\left(\frac{dp}{dV}\right)_S = c^2. \tag{1.4}$$

With increasing pressure, the speed of sound, as a rule, increases. An exception may be limited areas of abnormally high compressibility of a substance, which is usually associated with structural transformations, such as the dissociation of gases and liquids, polymorphic transformations of crystalline solids. For one-dimensional flows of compressible media, it is useful to enter the speed of sound in the substantial Lagrange coordinates:

$$a = \frac{V_0}{V}c = -V_0\left(\frac{\partial p}{\partial V}\right)_S^{1/2}. \tag{1.5}$$

1.2. Characteristics, Riemann integrals

In the case of a flat isentropic motion of a compressible medium, there are two families of characteristics – trajectories on a distance – time diagram, along which the set of partial differential equations (1.3) reduces to one differential equation of the first order. The C_+ and C_- characteristics correspond to the propagation paths of small perturbations (but do not necessarily describe the propagation of real perturbations) and are described in the Lagrange coordinates by the equations

$$\frac{dh}{dt} = a, \quad \frac{dh}{dt} = -a, \tag{1.6}$$

for the C_+ and C_- characteristics, respectively, that is, the slope of the characteristics is determined by the speed of the sound. It is important to emphasize that we can use the characteristics for any part of the diagram distance h – time t regardless of whether a change in pressure and density of a substance occurs at a given place at a given time or its state is kept constant.

Taking into account relations (1.2) and (1.5), the system of equations (1.3) along the characteristics is reduced to simple equations

$$du_p + \frac{1}{\rho_0 a} dp = 0 \quad \text{along } C_+ : \ \frac{dh}{dt} = a, \tag{1.7}$$

$$du_p - \frac{1}{\rho_0 a} dp = 0 \quad \text{along } C_- : \ \frac{dh}{dt} = -a. \tag{1.8}$$

The gas dynamics equations, written in the characteristic form (1.7), (1.8), are much more convenient for analyzing wave interactions than ordinary (1.3). Since the pressure p and the speed of sound a are uniquely related to each other by the equation of the state of matter, the differential expressions $du_p + dp/\rho_0 a$ and $du_p - dp/\rho_0 a$ are the full differentials of the values

$$J_+ = u_p + \int \frac{dp}{\rho_0 a} \quad \text{and} \quad J_- = u_p - \int \frac{dp}{\rho_0 a}, \tag{1.9}$$

which are called Riemann invariants. The Riemann invariant is constant along the characteristic and is determined with an accuracy of a constant, which is found from initial or other additional conditions. A more visual and convenient form of writing equations (1.9) is the relations between the particle velocity and pressure along the characteristics in the form of the Riemann integrals:

$$u_p = u_0 - \int_{p_0}^{p} \frac{dp}{\rho_0 a} \quad \text{along } C_+; \tag{1.10}$$

$$u_p = u_0 + \int_{p_0}^{p} \frac{dp}{\rho_0 a} \quad \text{along } C_-; \qquad (1.11)$$

where u_0, p_0 are the integration constants fixing the position of the state change trajectory along this characteristic on the $p - u_p$ plane. The Riemann integrals, which in some cases are not quite correctly called the Riemann isentropes, are two symmetric families of parallel curves in the coordinates pressure – particle velocity (Fig. 1.3). Product $\rho_0 a = \rho c = dp/du_p$ is the dynamic impedance of the substance. The concept of 'dynamic stiffness' is also used, by which is meant the product ρc^2.

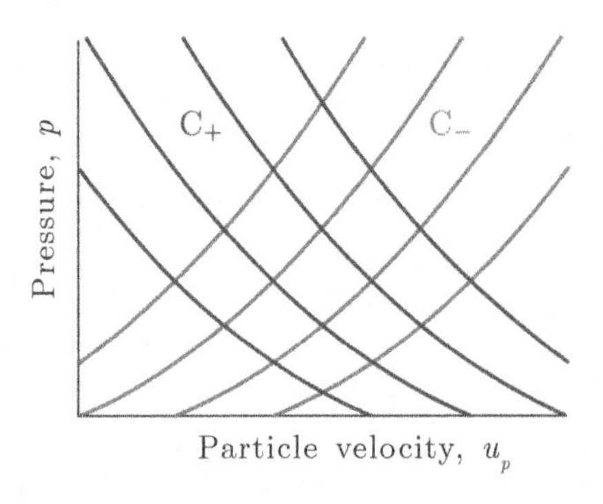

Fig. 1.3. Images of Riemann integrals on the particle velocity–pressure plane.

Figure 1.4 *a, b* illustrates an example of a flow during the propagation of a rarefaction wave in a uniformly compressed material at rest. At the time $t = t_0$, the left boundary of the compressed half-space was instantly released and the pressure on it abruptly dropped from the initial value p_i to zero. The pressure drop causes the appearance of a rarefaction wave, which propagates from the liberated boundary into the depth of the compressed substance. The forward front of the rarefaction wave moves with a velocity of the sound corresponding to the initial pressure p_i, however, the propagation velocity of subsequent sections of the rarefaction wave is less than the front velocity, since as the pressure drops, the sound velocity decreases. Due to the dependence of the speed of sound on pressure, the slope of the characteristics described in the *t–h* diagram the rarefaction wave increases as the corresponding pressure values decrease, as shown in Fig. 1.4 *a*, and the rarefaction wave itself expands as it propagates, as shown in Fig. 1.4 *b*. Since the

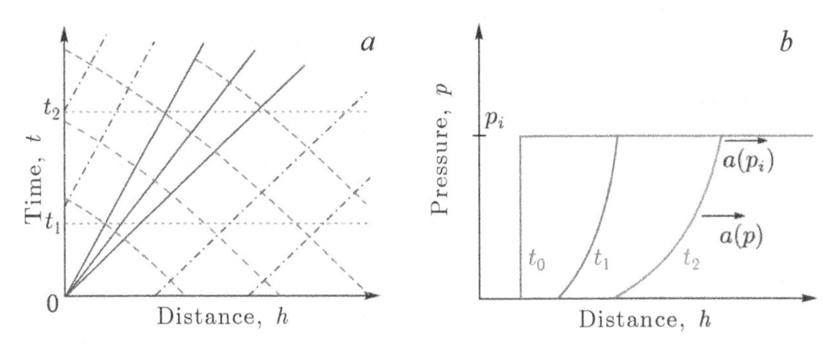

Fig. 1.4. Centred rarefaction wave with instantaneous release of the left boundary of a compressed half-space.

pressure was changed instantaneously, abruptly, at the boundary, all the characteristics describing the rarefaction wave emanate from one point with the spatial coordinate of the boundary of the compressed region and the time coordinate t_0. As a result, the set of characteristics describing a rarefaction wave on the *t–h* diagram takes the form of a fan. Such a wave, which in the *t–h* diagram is represented by a fan of characteristics emanating from a single pole, is called a centred rarefaction wave.

The motion of a compressible medium, in which all perturbations of states propagate in one direction, is a simple, or travelling wave. In a simple wave, the states along the characteristics directed towards the wave propagation are unchanged, and all states along any other trajectory on the *h–t* plane are described by a single dependence $p(u_p)$ corresponding to the Riemann invariant of opposite sign. The rarefaction wave shown in Fig. 1.4, is a simple centered wave. The characteristics of a simple wave are straight lines.

Figure 1.5 shows the distance – time and pressure – particle velocity diagrams for the case of the interaction of two symmetric counterpropagating rarefaction waves in a condensed substance. In this case, the flow consists of two simple waves everywhere except for the interaction region $A_1C_1C_3A_3A_1$. In this region, the values of pressure and particle velocity along the characteristics are variable and vary according to the corresponding Riemann integrals. At the intersection points of the characteristics, the values of pressure and particle velocity are the same for C_+- and C_-- characteristics and correspond to the intersection points of the Riemann isentropes describing the change of state along the intersecting C_+- and C_-- characteristics. In the figure, these states have the same designations in the *h–t* and $p–u_p$ diagrams. It can be seen from the figure that as

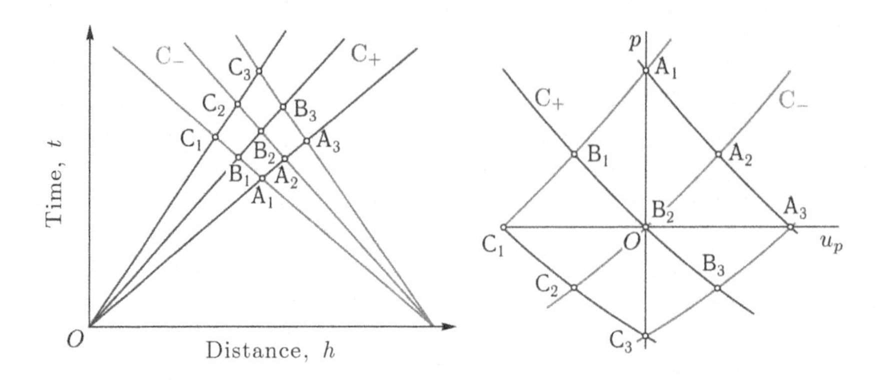

Fig. 1.5. The interaction of oncoming rarefaction waves. Equal indices indicate the states at the corresponding points of intersection of the characteristics.

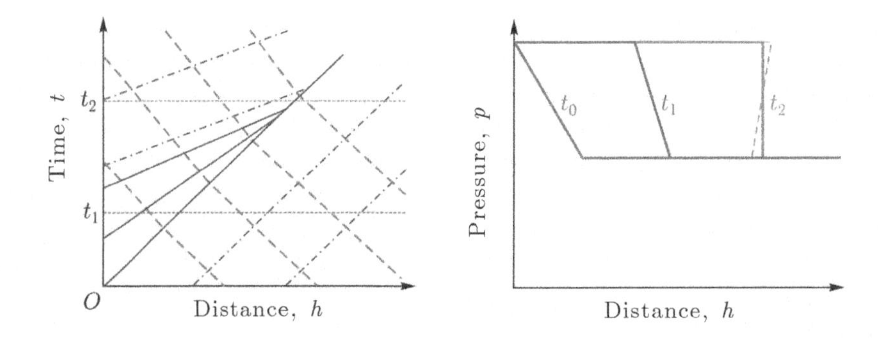

Fig. 1.6. The evolution of a 'ramped' compression wave as it propagates. In the distance–time diagram, dashed and dashed-dotted lines show, respectively, the characteristics of C_- and C_+ in a region free from disturbances; solid lines show a compression wave propagating in the positive direction. The pressure – distance diagram shows the formation of a shock wave from a ramped compression wave.

a result of the interaction of counterpropagating rarefaction waves in the body negative pressures are generated. This is possible only for condensed substances – solids and liquids. In gases, negative pressures are, of course, impossible; as the pressure approaches zero, the velocity of sound in a gas, in contrast to condensed substances, also tends to zero.

Figure 1.4 *b* shows plots of pressure $p(h)$ at different points in time. Due to the fact that the front of the rarefaction wave propagates with a velocity corresponding to the maximum pressure p_i, and the propagation velocity of the following sections of the rarefaction wave decreases as it is unloaded, the rarefaction wave expands as it propagates.

Figure 1.6 shows the evolution of a compression wave as it propagates. In this case, the dependence of the speed of sound on pressure leads to the fact that the steepness of the compression wave increases as it propagates. In normal environments where the speed of sound increases with increasing pressure, compression waves are transformed into shock waves, which in most cases can be interpreted as discontinuities or jumps in the parameters of the state of the medium. The system of differential equations of gas dynamics (1.3) cannot describe flows with discontinuities of state parameters.

1.3. Shock waves

The fundamental laws of conservation of mass, momentum, and energy of a substance in a shock wave are expressed by the Rankin–Hugoniot set of algebraic equations, which, if the pressure and velocity of matter are equal to zero in front of the wave front, have the form:

$$V = V_0 \frac{U_S - u_p}{U_S} \quad \text{(continuity equation)}, \tag{1.12}$$

$$p = \rho_0 U_S u_p \quad \text{(equation of conservation of the momentum)} \tag{1.13}$$

$$E - E_0 = -\frac{p(V - V_0)}{2} \quad \text{(equation of conservation of energy)} \tag{1.14}$$

where U_S is the velocity of propagation of a shock wave relative to an unperturbed medium with a specific volume V. As before, the system of equations is completed by the equation of state $E = E(p, V)$.

The derivation of equations (1.12), (1.13), (1.14) is illustrated in Fig. 1.7. Per unit of time, the shock wave travels a distance equal to U_S. During the same time, the particles of the cross section, in which the shock wave front was at the initial moment of time, shifted with the particle velocity to the distance u_p. As a result, the volume of a substance with a single cross-sectional area decreased from U_S to $U_S - u_p$. The same volume of matter with mass $m = \rho_0 U_S$ changed the velocity from zero to u_p, the corresponding change in the amount

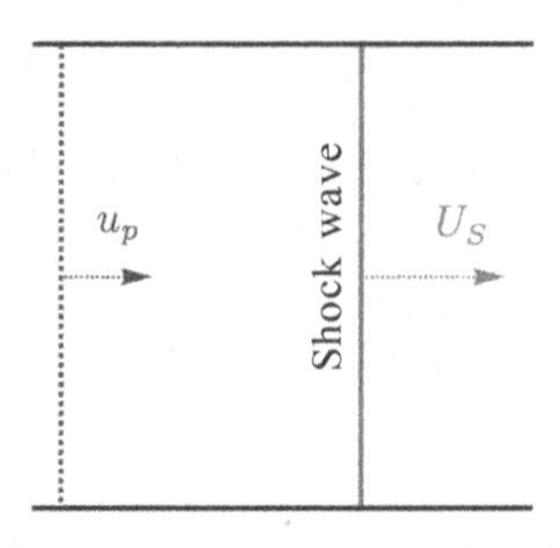

Fig. 1.7. To clarify the output of the conservation equations for a shock wave.

of movement under the action of a pulse of pressure forces p (with $\Delta t = 1$) is $\rho_0 U_S u_p$. Per unit of time of passage of a shock wave on the element of mass m, this element performs work pu_p, which is spent on the kinetic energy of this element $mu_p^2/2$ and changing its internal energy $m(E–E_0)$. If there is no energy release or absorption in the shock wave, then

$$pu_p = \rho_0 U_S \left(\frac{1}{2} u_p^2 + E - E_0 \right), \quad E - E_0 = \frac{pu_p}{\rho_0 U_S} - \frac{1}{2} u_p^2, \tag{1.15}$$

whence, taking into account (1.12), (1.13), we obtain (1.14).

Shock waves propagate at supersonic speeds relative to matter in front of the wave, but the velocity of the shock wave relative to the shock-compressed substance is subsonic – the disturbances in the shock-compressed matter catch up the shock wave front. The line connecting the initial state ahead of the shock wave front and the state p, V behind the shock wave is called the Rayleigh line or the Michelson line. Its equation is obtained by excluding the particle velocity from (1.12), (1.13):

$$p = \rho_0^2 U_S^2 (V_0 – V), \tag{1.16}$$

In the region of moderate shock pressures of condensed media the quasi-acoustic approximation works well, according to which the shock wave velocity is the arithmetic average of the speed of sound disturbances in front of and behind it:

$$U_S = [c_0 + a(p)]/2 = [c_0 + c(p) + u_p]/2. \tag{1.17}$$

The set of equations (1.12)–(1.14) together with the equation of state of matter determine its Hugoniot. The Hugoniot of a substance is the totality of its states that can be achieved as a result of shock-wave compression at certain fixed initial values of pressure and density.

In the range of moderate compressions, the Hugoniots of condensed media are usually described by a linear ratio of the form

$$U_S = c_0 + bu_p, \tag{1.18}$$

where the constant c_0 is equal to the speed of sound corresponding to the initial adiabateic volume compressibility of the substance, and b is a constant coefficient. Using the Rankin–Hugoniot equations, the corresponding values of pressure, specific volume, and energy are easily calculated from the given values of U_S and u_p. Figure 1.8 shows the Hugoniot of a substance in different coordinates.

According to equation (1.14), the increment of the specific energy of a substance in a shock wave is equal to the area of the

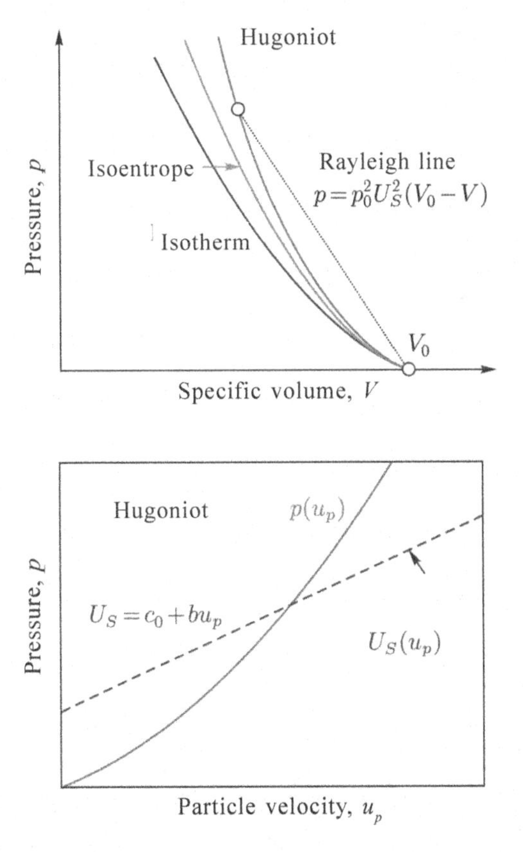

Fig. 1.8. Representation of the Hugoniot in different variables.

triangle under the Rayleigh line, while the increment of energy during isentropic compression to the same maximum pressure is equal to the area of the curvilinear triangle under the isentrope of the substance, which is obviously smaller. In other words, during shock compression, more energy is deposited in the substance and, accordingly, the substance is heated more than during the isentropic compression. For this reason, the Hugoniot in the p–V coordinates is located to the right of the isentrope.

Figure 1.9 shows the mutual position of the Hugoniot and two isentropes in the p–V and p–u_p coordinates. Shock compression is accompanied by an increase in entropy and, accordingly, excessive heating, so the Hugoniot in the p–V coordinates passes above the isentrope, drawn from a point with zero initial pressure and a specific volume V_0, but lower than the isentrope describing the unloading of the substance from the shock-compressed state. From the practice

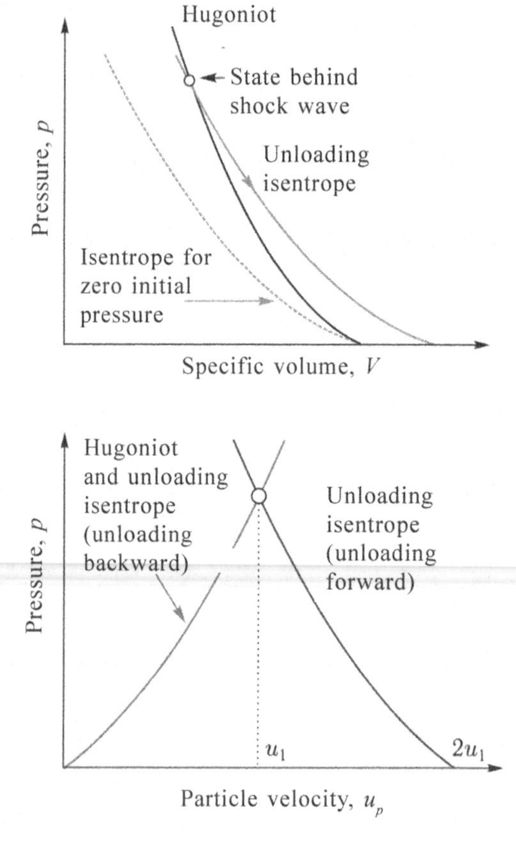

Fig. 1.9. The mutual position of the Hugoniot and isentropes in the p–V and p–u_p coordinates.

of shock-wave experiments with condensed substances, it is known that in the p–u_p coordinates, the Riemann isentropes of unloading coincide with good accuracy with the Hugoniot of the substance or are symmetric to it. It is interesting, however, to note that even in the case of exact coincidence in the p–u_p coordinates, in the p–V coordinates, the Hugoniot deviates very noticeably from the isentropes. The symmetry of the Hugoniot and isentropic unloading of the shock-compressed substance leads to the so-called doubling rule: the velocity of the free surface after the shock wave reaches it is equal to twice the particle velocity behind the incident shock wave. Doubling, of course, is approximate, but the error for non-porous condensed substances usually does not exceed 1–3%.

Another consequence of the coincidence of the Hugoniot and unloading isentrope is the possibility of estimating the speed of sound at high pressures. Along the isentrope it holds that

$$dp/du = \rho_0 a,$$

where a is the speed of sound in te Lagrange coordinates. Hence, for the Hugoniot in the form of (1.18), we obtain

$$a = c + 2bu_p,$$

which coincides with the quasi-acoustic approximation (1.17), or

$$a = \sqrt{c_0^2 + 4bp/\rho_0}. \qquad (1.20)$$

Under stepwise compression, the Hugoniot of a substance compressed by the first shock wave, in the p–u_p coordinates, usually almost coincides with the initial Hugoniot. However, in the p–V coordinates, the Hugoniot of the second compression is always located to the left of the initial Hugoniot. The reasons for this are illustrated in Fig. 1.10. Suppose that single and stepwise shock compression are described by a single Hugoniot OH_1H. But in this case, when compressing up to the same pressure, the p_H increment of internal energy in one shock wave, equal to the area of the OHC triangle, is larger than the increment of internal energy during step compression by two shock waves, equal to the sum of the areas of the OH_1C_1 triangle and C_1H_1HC trapezium. For this reason, due to different thermal expansion of the state, realized in the second shock wave, it should be characterized by a smaller specific volume, as

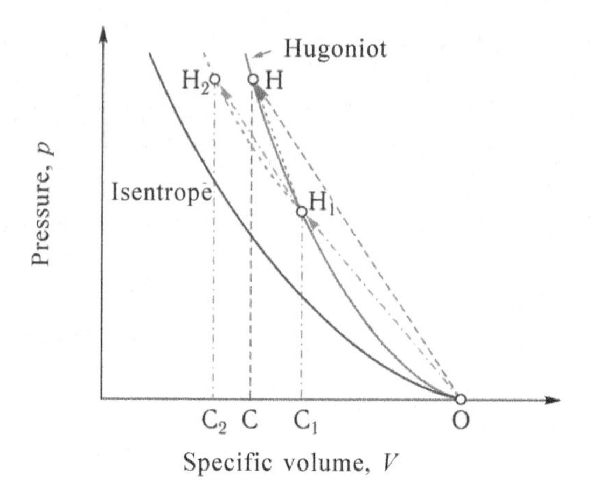

Fig. 1.10. Stepwise compression in series with two shock waves.

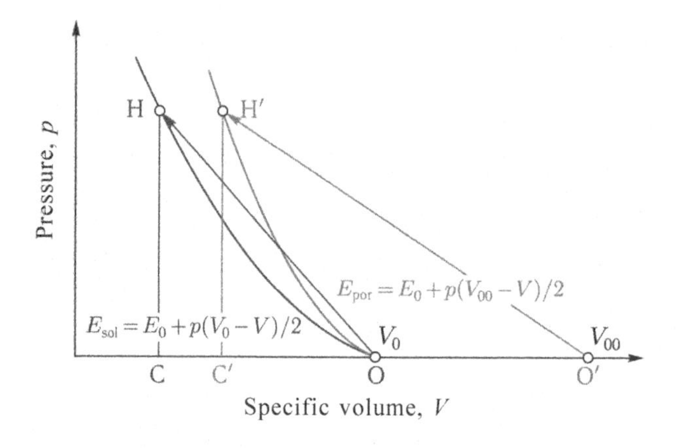

Fig. 1.11. Shock compression of porous media.

shown by the adiabateic secondary compression H_1H_2 in Fig. 1.10. During multi-stage compression of a substance, a series of successive shock waves realizes states close to the isentrope. In this case, one speaks of quasi-isentropic shock compression.

Higher temperature states are realized under shock compression of low-density samples made of powder materials or 'foams'. Indeed, since the volume change during compression of a porous substance involves closing the pores, then at the same pressure of the shock compression p_H, the energy increment in the porous substance with the initial specific volume V_{00}, equal to $E_{por} - E_{00} = p_H(V_{00} - V)/2$ (Fig. 1.11) turns out to be larger than the energy increment in shock compression of a solid substance, equal to $E_{sol} - E_0 = p_H (V_0 - V)/2$,

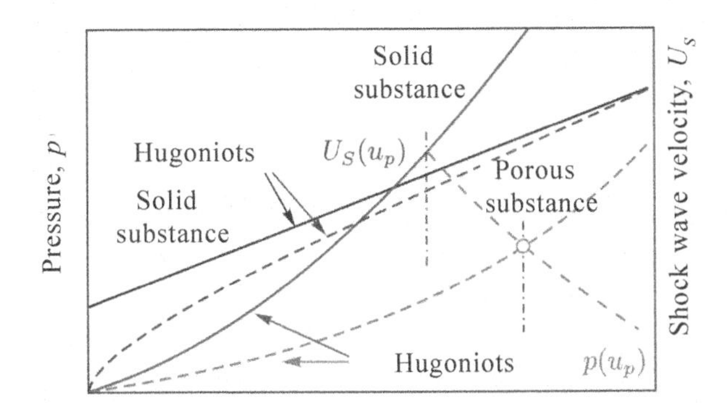

Fig. 1.12. The mutual position of the Hugoniots of solid (solid lines) and porous (dashed lines) media in the particle velocity–velocity coordinates of the shock wave and particle velocity–pressure coordinates.

$V_{00} > V_0$. Excess energy leads to thermal expansion of the substance, the Hugoniot of the porous substance (OH' in Fig. 1.11) is located to the right of the Hugoniot of solid matter (OH). The difference in the values of the initial internal energy E_{00} and E_0 is equal to the surface energy of the grains of the porous substance and is insignificant when compared with the jump in energy in a strong shock wave. The use of porous materials to expand the temperature range of states attainable in experiments with shock waves was proposed by Ya.B. Zel'dovich and is widely used in the practice of measuring parameters of equations of state of condensed media.

Figure 1.12 compares the Hugoniots of a substance in solid and porous states in the particle velocity–velocity of the shock wave and particle velocity–pressure coordinates. In the $u_p - U_S$ coordinates, the Hugoniot of a porous medium is substantially nonlinear, located below the Hugoniot of solid matter and should, generally speaking, proceed from zero coordinates. In the $u_p - p$ coordinates, the Hugoniot of the porous substance is also located below the Hugoniot of the same substance in a continuous, non-porous initial state. At the same time, while for describing the compression and rarefaction waves propagating in the shock-compressed solid substance in these coordinates, we can use the single dependence – its Hugoniot, for the porous substance this approximation is no longer acceptable. After compression in the first shock wave, we are already dealing with a non-porous substance with a corresponding compressibility.

1.4. Stability of shock waves, multiwave configurations and rarefaction shock waves in a medium with abnormal compressibility

Shock waves propagate at supersonic speeds relative to matter in front of the wave, but disturbances in the shock-compressed matter catch up the shock wave front. The fact that a shock wave is subsonic with respect to a shock compressed substance ensures its stability. Compression waves are transformed into shock discontinuities due to the fact that the speed of sound increases with pressure, as is usually the case. However, there are materials whose compressibility varies with pressure in a non-monotonic manner, as shown in Fig. 1.13.

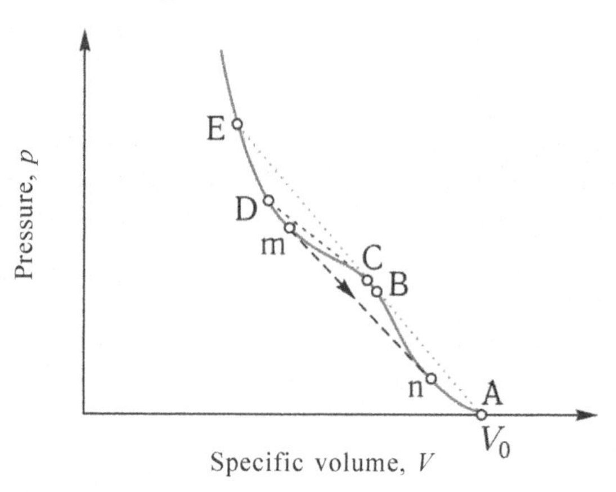

Fig. 1.13. An example of abnormal compressibility. For ease of illustration, the Hugoniot and isentrope are represented by a single curve.

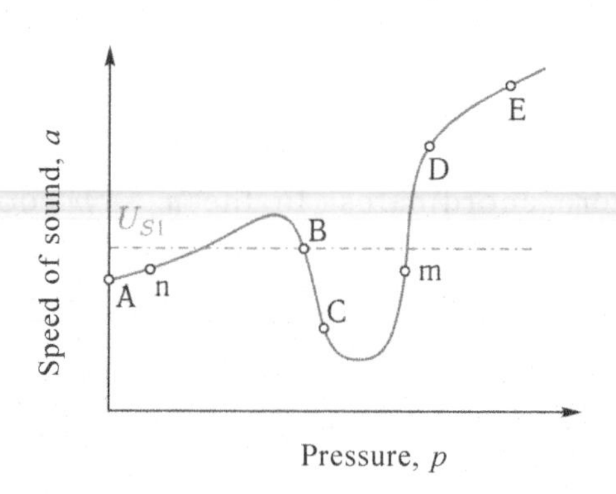

Fig. 1.14. The dependence of the speed of sound on the pressure in a substance with abnormal compressibility. Designations are the same as in Fig. 1.13. U_{S1} is the speed of the first shock wave AB.

Figure 1.14 shows for this material the dependence of the velocity of propagation of perturbations on pressure. If we consider the evolution of a 'smooth' compression wave with a final pressure p, $p_B < p < p_E$ in such material, we will see that the steepness of its initial section increases with pressure from zero to p_B as it propagates. Starting from point B, the velocity of propagation of perturbations in a compressed substance is lower than the velocity of the shock wave being formed; these perturbations are lagging behind and do not participate in the formation of a shock jump. In the region of higher pressures, the perturbation speed increases again as it is compressed; therefore, the formation of a shock wave in this region again becomes possible. As a result, a smooth compression wave evolves into two shock waves AB and CD (Fig. 1.15), propagating one after another at different speeds, and an expanding region of smooth growth BC between them. The BCDE portion of the Hugoniot in Fig. 1.13 is called a 'closed area'. The states of a substance corresponding to this region cannot be reached by a single shock jump from the initial state A. With an increase in the pressure of shock compression to the value of p_E corresponding to the intersection point of extrapolation of the Rayleigh line AB with the upper branch of the Hugoniot (Fig. 1.13), the velocity of the second shock wave becomes equal to the velocity of the first wave and the two-wave configuration disappears. When unloading a compressed substance, a non-monotonic change in the speed of sound with decreasing pressure leads to the formation of a shock rarefaction wave *m–n* (Fig. 1.15).

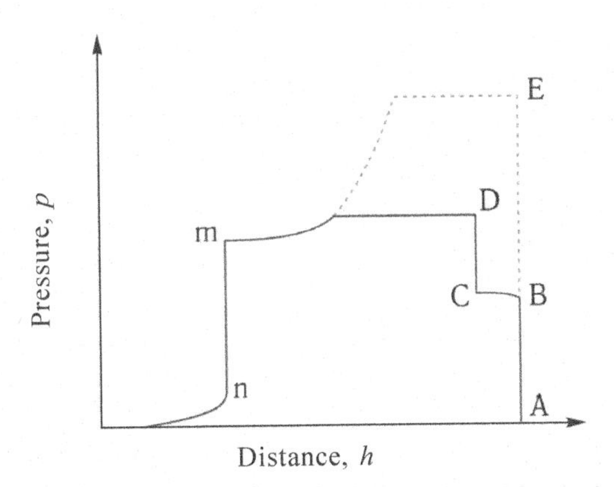

Fig. 1.15. Compression pulse in a substance with abnormal compressibility. The designations correspond to Figs. 1.13 and 1.14.

1.5. Decomposition of discontinuities and wave interactions

Shock waves and simple Riemann waves constitute an important class of self-similar flows on which dynamic methods for studying equations of state of matter are based. In this case, the diagnostics of measured states is based on the solution of the problem of the decomposition of an arbitrary discontinuity. The solution of the problem of the breakup of a discontinuity is a combination of shock waves and centered rarefaction waves propagating from the place of the initial discontinuity and separated by the region of constancy of the state parameters. The analysis of decays of discontinuities and other wave interactions is a necessary element of experiments with shock waves.

In problems on the decomposition of a discontinuity, we will be interested in two typical cases: 1 – shock waves propagate to both sides of the discontinuity; 2 – in one direction the shock wave moves, and in the other – a centred rarefaction wave. The corresponding pressure profiles and phase trajectories of the processes in the $p - u_p$ diagrams are presented in Figs. 1.16 and 1.17. The situation with two shock waves takes place, for example, when plates collide or when a shock wave is reflected from the interface with a substance having a higher dynamic impedance ρ_c. The second case is realized when a shock wave is reflected from the interface with a less rigid medium (i.e., a medium having a lower dynamic impedance).

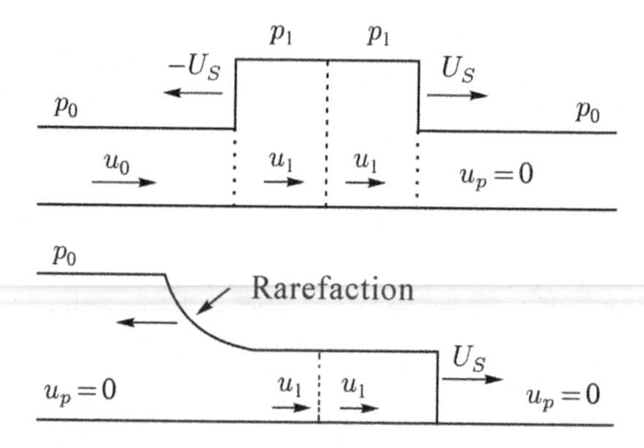

Fig. 1.16. Pressure plots formed as a result of decomposition of discontinuities in the distributions of particle velocity (*a*) and pressure (*b*). The vertical dashed lines indicate the initial positions of the discontinuities, the arrows in the lower part indicate the directions of motion of the substance, the arrows above show the directions of wave propagation.

Analysis of discontinuity decays is usually carried out by parallel construction of the $t - x$ and $p - u_p$ diagrams. Waves of compression and rarefaction, resulting from the decomposition of the discontinuity, should transfer the substance on both sides of its initial position to states with equal values of pressure p and particle velocity u_p. Consequently, the desired values of p, u_p must simultaneously satisfy two dependences $p(u_p)$, which describe the state changes for waves of positive and negative directions. When analyzing discontinuity decomposition, it should be taken into account that for positively directed waves the slope of the phase trajectories is $\partial p/\partial u_p > 0$, and for waves moving in the negative direction, $\partial p/\partial u_p < 0$.

Two shock waves are formed as a result of the decomposition of the discontinuity in the particle velocities (Fig. 1.16 *a*) from 0 to u_0; one of these waves propagates in the positive direction, and the other in the negative. The states of shock-compressed matter behind these waves should satisfy the corresponding Hugoniots in Fig. 1.17 *a*. For a wave of positive direction, the Hugoniot in coordinates p, u_p has a positive slope and passes through the point $p = p_0$, $u_p = u_0$, denoting the initial state of the particles of matter to the right of the discontinuity. For a wave of negative direction ($U_S < 0$), the Hugoniot has a negative slope and passes through the point $p = p_0$, $u_p = 0$, which describes the initial state of the particles to the left of the discontinuity. The condition of equal pressures and particles velocities is satisfied at the point $p = p_1$, $u_p = u_1$, the intersection of these two Hugoniots.

A shock wave and a rarefaction wave are formed as a result of the decomposition of the discontinuity in pressures from 0 to p_0 (Fig. 1.16 *0*). In the example shown, the shock wave propagates into the uncompressed substance in the positive direction, and the rarefaction wave propagates through the compressed substance in the negative direction. This rarefaction wave is simple; therefore, the process of unloading compressed matter is described by a Riemann isentrope with a negative slope passing through the point of the initial state $p = p_0$, $u_p = 0$. The result of the decomposition of the discontinuity corresponds to the point of intersection of the Hugoniot and the Riemann isentrope in the coordinates p, u_p (Fig. 1.17 *b*).

1.6. Detonation wave

Detonation is the process of chemical transformation of explosive substances (ES) that occurs in a very thin layer and propagates at

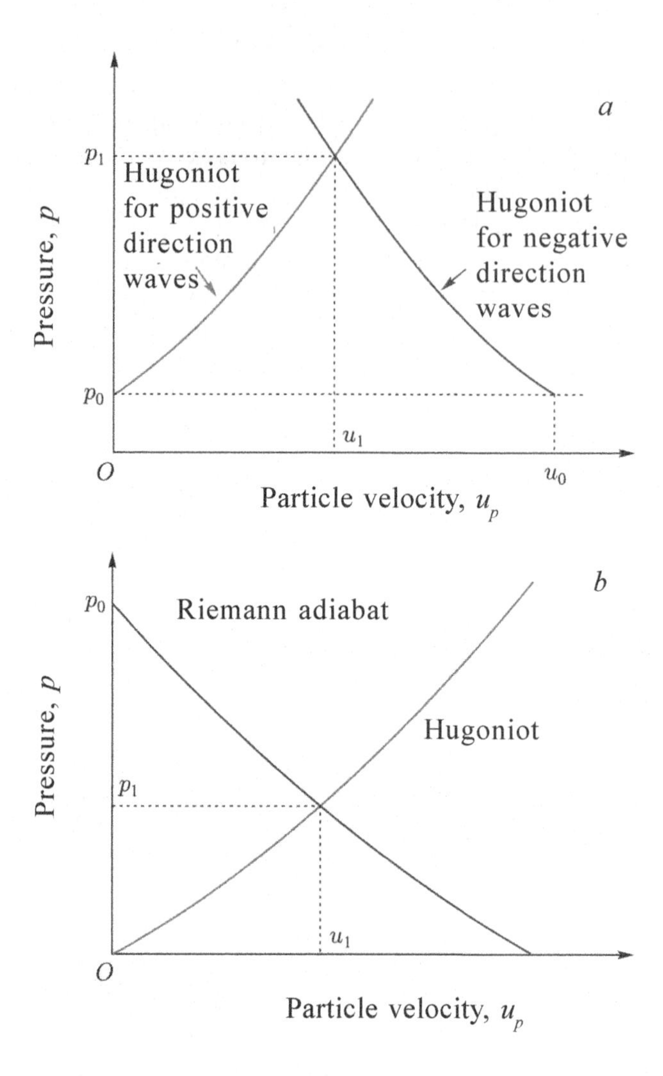

Fig. 1.17. The particle velocity–pressure diagrams for the decomposition of discontinuities shown in Fig. 1.16.

supersonic speeds. In accordance with the theory of Ya.B. Zel'dovich, the detonation transformation of explosives is carried out under the action of a shock wave, which, propagating in the charge, excites an exothermic chemical reaction. In turn, part of the released energy is spent on maintaining the further movement of the shock wave. There is a stationary speed of self-sustaining detonation, which does not depend on the initiating pulse and is determined only by the energy released during a chemical reaction (heat of explosion) and the equation of state of the detonation products.

For the detonation wave, the same equations of mass and momentum conservation are valid as for the shock wave. The energy conservation equation (1.14) must take into account the release of energy as a result of a chemical reaction:

$$E - E_0 = q' - \frac{p(V - V_0)}{2}, \tag{1.21}$$

where q' is the caloric content of HE (the amount of released energy per unit mass). If the equation of state of the explosion products $E(p, V)$ is known, then equation (1.21) allows us to construct in the plane (p, V) a curve of possible states behind the detonation wave – the so-called detonation adiabat.

The structure of a plane detonation wave is shown in Fig. 1.18. According to the theory of Ya.B. Zel'dovich, the detonation complex includes a shock wave, a chemical reaction zone (chemical spike) of constant width, and an unsteady region of expanding explosion products. Chemical reactions require for their completion a certain finite time determined by their kinetics. The rate of chemical reactions increases exponentially with temperature. Compression and adiabatic heating of a substance in a shock wave occur extremely quickly, chemical reactions do not have time to occur during the shock compression, so the state of the particles of matter immediately behind the shock jump corresponds to the Hugoniot of unreacted explosives. The shock wave initiates exothermic chemical reactions in explosives, and the composition and temperature of the shock-compressed substance change. In the process of energy release and the formation of products of chemical transformation, the state of matter deviates from the Hugoniot of unreacted explosives. From the condition of the constancy of the velocity of a steady detonation wave, it follows that changes in pressure and specific volume of a substance in the reaction zone should occur along the Rayleigh line $p = \rho_0^2 U_S^2 (V_0 - V)$.

At the end of the reaction zone, the condition of equality of the flow speed of a substance relative to the detonation wave front and the local Lagrangian speed of sound is satisfied (Chapman–Jouguet condition). On the p–V graph, the Chapman–Jouguet (Ch–J) condition requires that the Rayleigh line describing the change in the explosive state in the reaction zone be related to the isentropic explosion products (EP). At the same time, the detonation adiabat is touched. The motion of a substance in the region of expansion of the explosion products is supersonic with respect to the detonation

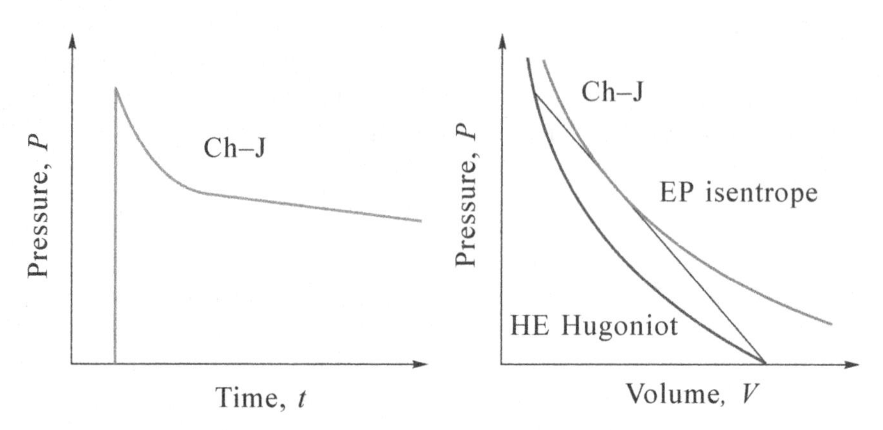

Fig. 1.18. The structure of a plane detonation wave and a change in the state of an explosive in a chemical reaction zone. ChJ – Chapman–Jouguet point.

wave; therefore, perturbations from this region do not penetrate into the reaction zone and cannot influence the detonation velocity. A subsonic flow takes place inside the reaction zone, which ensures the transfer of the released energy of chemical transformation to the shock wave front. The Chapman–Jouguet determines the propagation velocity of a plane stationary detonation wave D. In other words, the detonation velocity is completely determined by the equation of state of the explosion products, the heat of the explosion and the initial density of the high explosive. The point of tangency of the Rayleigh line with the EP isentrope on the p–V graph is called the Chapman–Jouguet point, and the plane in the one-dimensional detonation wave at which the velocity of propagation of the disturbances becomes equal to the detonation velocity is called the Chapman–Jouguet plane. The detonation speed of condensed explosives is several kilometers per second – up to 9 km/s in the case of, for example, HMX, and varies depending not only on the chemical composition, but also on the density.

There is a critical diameter of the explosive charge, such that charges of a smaller diameter are incapable of self-sustaining detonation. An explanation for this phenomenon was given by Yu.B. Khariton on the basis of the fact that a certain time is needed to complete the energy release in the detonation wave. According to the Khariton principle, the critical conditions for detonation are determined by the equality of the reaction time of a compressed substance and the time of its lateral expansion. In turn, the time to the expansion is determined by the radius of the explosive charge and the speed of sound in it. The velocity and pressure of detonation decrease with the approach of the diameter of the charge of a high-explosive to the failure value.

2

Methods of generation of shock waves and measurement of gas-dynamic parameters in dynamic experiments

In this chapter, we will briefly discuss the methods for exciting and recording shock loads in condensed media, which are used to obtain information on the compressibility of a substance and its mechanical and kinetic properties. A more detailed description of the experimental technique and the corresponding bibliography can be found in [3, 4].

Active research in the field of shock wave physics was started during the Second World War with the aim of obtaining thermodynamic equations of the state of condensed media in a wide range of pressures and temperatures. To carry out the necessary measurements of the shock compressibility of substances during this period, explosive generators of plane shock waves were created, and discrete methods were developed for measuring the speed of shock waves and the velocity of movement of the sample surface. The logic of the further development of experimental technology led to the development of methods for the continuous monitoring of pressure and particle velocity in whole shock load pulses, which opened up new possibilities for studying the mechanical and kinetic properties of various materials and chemically active substances under conditions of shock-wave loading. The radical improvement in the spatial and temporal resolution of modern measurement methods made it possible to study extreme states in the laboratory using advanced intensive pulse load generators, such as lasers, relativistic electron and ion beams.

2.1. Explosive generators of plane shock waves

The simplest way to excite a shock wave in a solid with an amplitude of several tens of gigapascals is to explode the charge of a high explosive (HE) on the sample surface. For ease of interpretation of the measurement results, it is desirable to have a flat stationary shock wave in the sample. Plane shock and detonation waves are formed using various flat-wave oscillators.

Most often, conic explosive lenses are used as plane-wave generators (Fig. 2.1), whose principle of operation is based on the use of elements with different detonation speeds. Such explosive lenses consist of an external conical explosive charge with a high detonation velocity and an inner explosive charge with a low detonation velocity (usually a mixture of barium nitrate with TNT – barathol) or inert material (usually lead or paraffin). High-speed detonation in the outer cone excites in the liner a detonation or shock wave inclined to form a cone at an angle φ defined by the ratio of wave velocities in the outer and inner elements: $\sin \varphi = D_{liner}/D_{end}$. The aperture angle of the cone ψ is chosen such that a plane wave emerges at the base of the liner: $\psi = \pi - \varphi$.

A pressure drop begins due to the explosion of the explosive products in the detonation wave immediately behind the shock jump. For stationarity of the shock wave process, it is necessary to form a region of constant parameters behind the shock wave. Impulses of

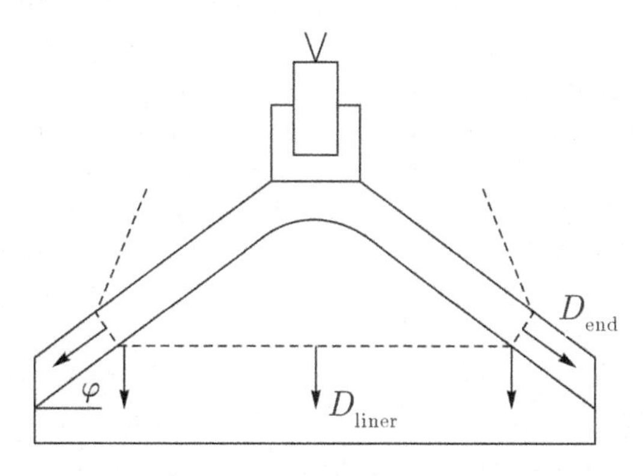

Fig. 2.1. Explosive lens.

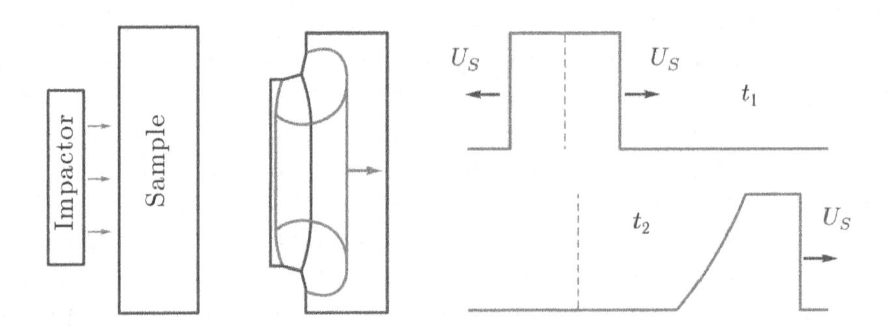

Fig. 2.2. Generation of a compression pulse by a plate. The shock waves propagating from the impact surface and rarefaction waves propagating from the back surface of the impactor and the lateral surfaces are shown.

a shock load with a constant, for some time, pressure after a shock, are generated by the impact of a flyer plate. The sufficiently large transverse dimensions of the impactor and the sample ensure the one-dimensional motion of the medium during the period of time required for measurements. The one-dimensionality of the process is disturbed by unloading waves propagating from the lateral surfaces of the impactor and the sample to the centre (Fig. 2.2). The longer the required recording time, the larger should be the diameter of the flat part of the impactor.

At the moment of impact, shock waves diverging from the impact surface are excited in the impactor and the sample. The parameters of the shock wave in the sample are determined from the condition of equality of pressures and mass velocities on both sides of the impact surface and their correspondence to the Hugoniots of the impactor and the sample. The desired parameters correspond to the intersection point in the coordinates p, u of the Hugoniot of the sample material and the Hugoniot of braking the material of the impactor. A shock wave is reflected on the back surface of the impactor. A rarefaction wave appears, propagating through the sample at a speed of sound. During the time of reverberation of the waves in the impactor, a constant pressure is maintained on the impact surface. The one-dimensionality of the process on the axis of impact is maintained until the lateral rarefaction wave converges. This determines the minimum ratio of the diameter of the impactor d to its thickness δ. Usually, when registering a complete wave profile (a shock wave and a rarefaction wave following it), the d/δ ratio is taken much more than 5. Similar requirements are imposed on the sample size.

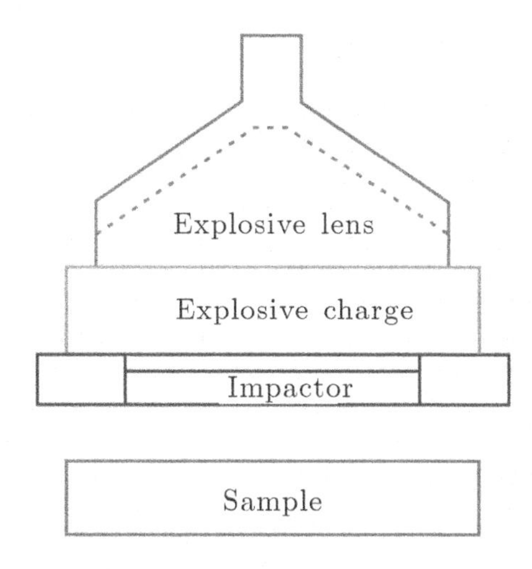

Fig. 2.3. Scheme of an explosive lauching device,

Launching of impactors is carried out by detonation of charges of explosive or by using powder or gas guns or other devices. Figure 2.3 shows a typical scheme of explosive facilities, widely used in experimental physics of high dynamic pressures. Such devices make it possible to accelerate metal or plastic impactors with a thickness of 1–10 mm to speeds of about 1–6 km/s. The impactor maintains its flat surface in the central part despite the fact that due to the radial expansion of the explosion products, the pressure at the periphery of the charge drops faster than at its axis. The correction of the pressure pulse acting on the edge of the impactor is achieved by using a focusing ring, which causes a rise in pressure at the periphery of the charge upon reflection of the detonation wave and additional inleakage of the products of the explosion into the cavity above the impactor. The cavity also serves to 'mitigate' the loading process of the impactor and prevent its spallation fracture. For the given mass of the explosive charge and the transverse dimensions of the impactor, the speed of launching is greater, the smaller the thickness and density of the flyer plate. The natural limitation on the throwing speed is the speed of expansion of the explosion products into the atmosphere, which is approximately equal to the detonation velocity.

One of the advantages of the explosive launching of the flyer plates – impactors is the possibility of organizing a symmetric impact, in which two shock waves are simultaneously generated in the sample, propagating towards each other. The interaction of the

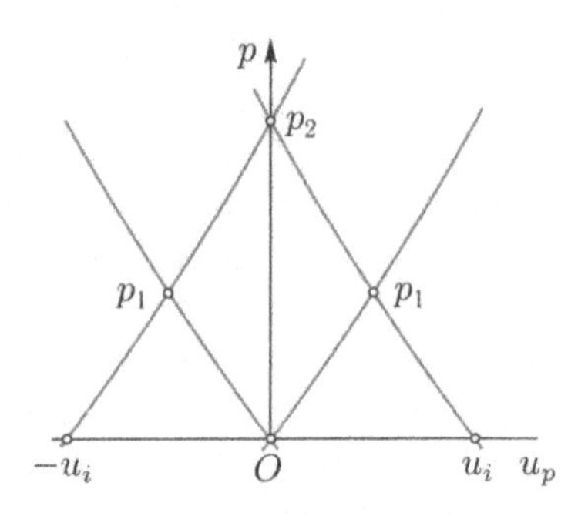

Fig. 2.4. An explosive device for the generation of oncoming shock waves and a wave interaction scheme for a counter impact.

counterpropagating shock waves makes it possible to achieve higher pressures while maintaining a relatively long duration of their action. A photograph of such an experimental device and a wave interaction scheme with a counter impact are shown in Fig. 2.4. The impact of plates with velocities u_i and $-u_i$ in the target generates shock waves with pressure p_1, propagating towards each other. After their collision, shock waves with a pressure jump from p_1 to p_2 are formed, which propagate in the shock-compressed substance from the plane of their meeting to the impact surfaces. The pressure p_2 behind these waves more than doubles the pressure p_1.

Devices with counter-impact are used to measure the electrical conductivity of various substances depending on pressure. In this case, the test substance in the form of a foil or film is placed between polymer insulating gaskets and, together with them, is clamped between plates of high-impedance material – usually copper or sapphire. Under these conditions, multiple reflections of the waves occur in the low-impedance gasket, leading to a stepwise increase in pressure, as shown in Fig. 2.5. The process ends with the fact that the pressure in the gasket is set equal to the pressure of shock compression of the main plates. The placement of the test substance in the form of a layer between the plates of a high-impedance material is used to study its properties under conditions of quasi-isentropic compression. In such substances as liquid hydrogen and other gases, it is very difficult to produce pressures of the megabar range by impact, and this formulation of experiments allows to solve the problem.

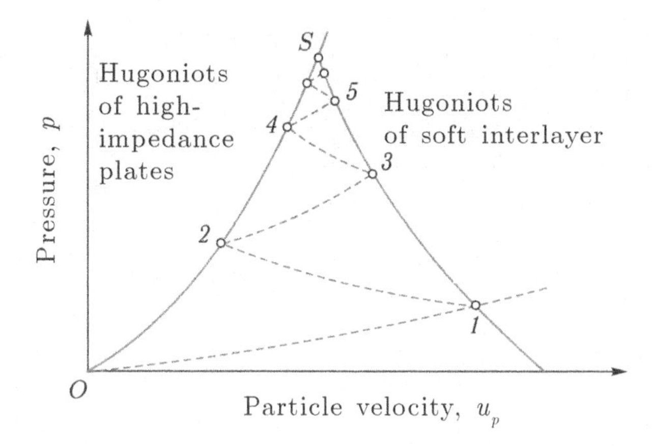

Fig. 2.5. Multiple reflections of waves in a thin gasket between the plates of high-impedance material.

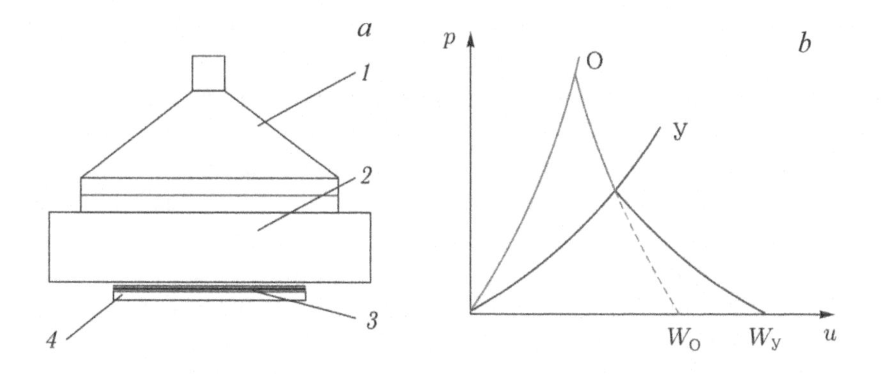

Fig. 2.6. Low-speed launching of flyer plates by a shock wave. *a* – explosive device scheme: 1 – explosive lens; 2 – plate attenuator; 3 – 'shock absorbing' gasket made of material with low dynamic impedance; 4 – flyer plate. *b* – diagram of wave interactions: O, Y – Hugoniots of the attenuator plate and the impactor; W_O, W_Y - surface speeds of the attenuator plate and the impactor.

While maintaining direct contact between the impactor and the explosion products, it is difficult to get the flyer plate speed less than 1 km/s. To reduce the speed, plate attenuators are used from a material with a higher dynamic impedance than that of the impactor. A scheme of such a device and a diagram of wave interactions are shown in Fig. 2.6. An explosive lens creates a shock wave in the attenuator plate with the parameters corresponding to the '*O*' point on the *p*, *u*-diagram. Due to the difference in the dynamic impedances of the materials of the attenuator and the impactor, the latter, when the

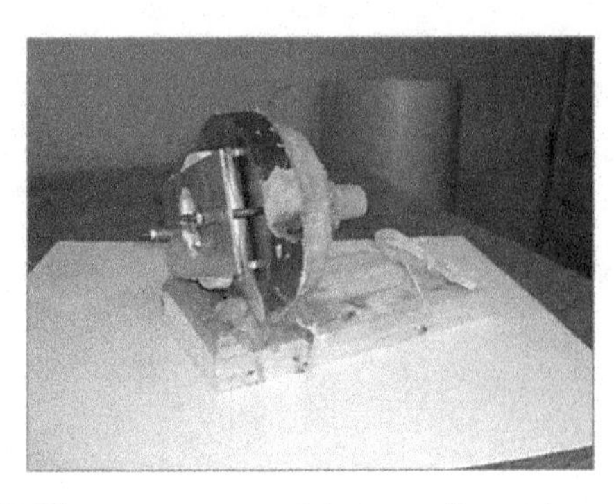

Fig. 2.7. The appearance of the experimental assembly.

shock wave reaches its back surface, acquires a speed higher than the attenuator. As a result, the impactor is separated from the attenuator.

Figure 2.7 shows an example of an experimental assembly with a steel attenuator for low-speed launching of a flat aluminium impactor.

2.2. Ballistic facilities for experiments with shock waves

Explosive shock wave generators are compact, cheap and provide a wide range of amplitudes and durations of load pulses. However, the destructive effect of an explosion imposes specific safety requirements. To work with such devices, special expensive explosive chambers or open test areas are needed, as well as technological equipment for the remote production of high-quality profiled charges. An alternative solution is the use of smooth-bore ballistic installations. The indisputable advantage of such installations is the possibility of smooth adjustment of the velocity of impact, ensuring minimal and controlled distortion of the impactor relative to the sample plane in each experiment, high uniformity of the region of one-dimensional flow behind the shock wave front in the sample. It is also important for measurements of shock compressibility that the impactor practically does not undergo heating during acceleration, as is the case with the use of explosive devices.

To study the mechanical properties of materials under shock-wave loading, gas guns with a caliber of 25–100 mm are usually used. With a barrel length of up to 14 m and an initial pressure of compressed gas (nitrogen or helium) up to 150 atm, such installations produce launching speeds in the range from 100 to 1000 m/s. The launching speeds of powder guns are 400–2500 m/s. A flat impactor is installed on the end of a hollow cylindrical projectile. To achieve the greatest parallelism of the colliding planes, the target is placed in such a way that the impact occurs at the moment when the projectile was not yet completely out of the barrel. The method allows the installation of the sample and sensors both on the target and on the projectile itself.

From ballistics it is known that the maximum velocity of the projectile w is determined by the formula:

$$w = c_g \sqrt{2f / \gamma(\gamma - 1)(m_p / m_g + b_1)}, \tag{2.1}$$

where c_g is the speed of sound in the gas pushing the projectile, f is the thermal efficiency in the process of gas expansion, γ is the adiabatic exponent (ratio of specific heats), m_p is the mass of the projectile, m_g is the gas mass, b_1 is a constant (the proportionality coefficient between the velocity of the projectile and the average velocity of the gas). This ratio shows that the launching speed is limited by the value proportional to the speed of sound in the pushing gas, and does not exceed it. The speed of sound in a gas increases with increasing pressure and temperature and decreasing of its molecular weight. Therefore, in order to increase the throwing speed, it is necessary to increase the temperature and pressure in the pushing gas and reduce its molecular weight.

The highest launching speeds are achieved in multi-stage light-gas installations. The task of the extra steps is to heat and compress the light pushing gas. Figure 2.8 is a scheme of a two-stage gun. Chamber 4 contains light gas, which is compressed and adiabatically heated by piston 2 driven by powder gases 1. After the pressure in the light gas chamber reaches a predetermined value controlled by the strength between the light gas chamber and the second stage, the projectile 5 begins to move second stage barrel 6. The initial light gas pressure is usually 10–100 atm, and at the time of the shot it can reach 10 000 atm. The maximum throwing speed obtained on a two-stage light-gas installation reaches the value of the second

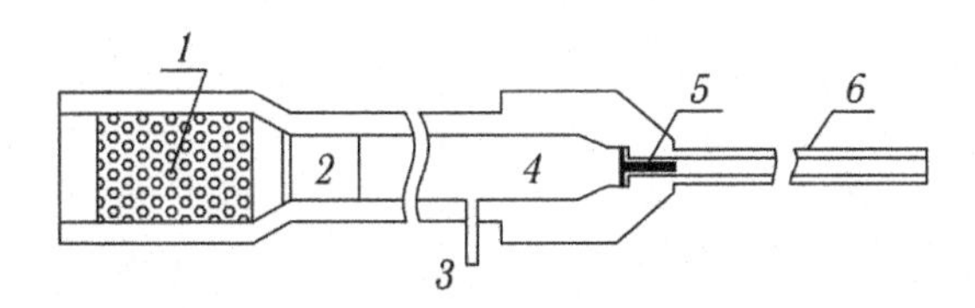

Fig. 2.8. Scheme of a two-stage ballistic installation.

Fig. 2.9. Two-stage light gas gun for shock wave experiments at the Sandia Laboratories (USA).

space velocity of 11.2 km/s with a projectile mass of 0.04 g and a caliber of 5.6 mm. For experiments with plane shock waves, two-stage light-gas installations with a barrel diameter of 12–25 mm and a length of up to 50 m are used, at which throwing speeds of up to 8 km/s are obtained.

Figure 2.9 shows a two-stage light-gas gun for shock wave experiments at the Sandia Laboratory (USA). Ballistic installations designed for shock-wave measurements are usually equipped with a sensor system for measuring the velocity and relative tilt of the projectile in each experiment, accurate synchronization of the measuring equipment and determining the possible distortion of the surface of the impactor during acceleration. The barrel in front of the projectile and the space around the sample are evacuated. In general, a two-stage ballistic installation is a rather complicated and expensive construction with a long cycle of preparation and conducting experiments and restoring readiness for further work.

2.3. Prospective high dynamic pressure generators

Explosive and ballistic generators of plane shock waves are now the main tools in high dynamic pressure physics. The desire to further advance into the region of previously inaccessible parameters and organization of shock-wave research in the context of a conventional physical laboratory encourages the use of new, non-traditional methods of intensive dynamic effects on matter. It is proposed to use electric explosive devices, high-power pulsed lasers and high-current electron and ion accelerators as promising sources of high dynamic pressures.

An electrical explosion of conductors that occurs when a capacitor battery is discharged onto them is used to throw plates about 0.1 mm thick. In the latter case, the dense plasma of a flat conductor (exploding foil) resulting from an electric explosion pushes the piston of a dielectric material, which can be covered by an additional thin impactor coil of a material with high dynamic impedance. By varying the voltage and capacitance of a capacitor battery, the size of an exploding foil, it is possible to change the specific energy of an electric explosion over a wide range and thereby vary the throwing speed of thin impactors from 10–100 m/s to 18 km/s and more.

In order to generate compression pulses, a number of studies use high-power pulses of laser or particulate radiation. In order to understand the mechanism of such a method of generating compression waves, let us analyze the mechanical action of instantaneous volume heat deposition, which occurs, for example, when an intense pulse of penetrating radiation is applied to a solid target (Fig. 2.10).

Absorption of radiation in the near-surface target layers causes an increase in the temperature of these layers. If the duration of exposure is sufficiently short so that the movement of the target surfaces does not have time to noticeably affect the process of interaction of radiation with the substance, then the heating of the absorbing layer of the target occurs almost at constant volume and is accompanied by an increase in pressure. As the distance *h* in the

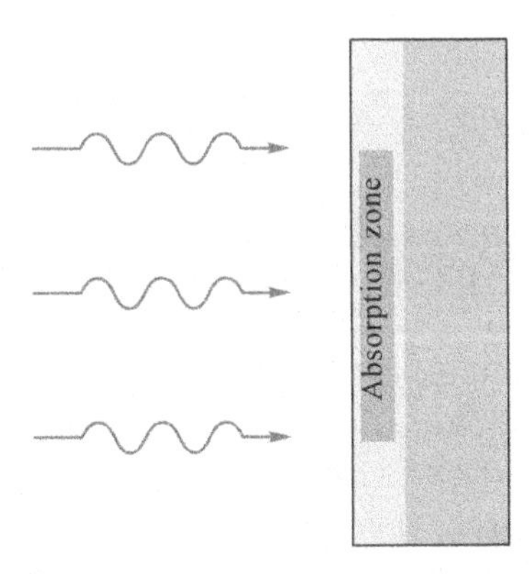

Fig. 2.10. Impact of a pulse of penetrating radiation on a solid target.

target from its surface increases, the value of the absorbed energy and, accordingly, the pressure changes. For the sake of simplicity, let us assume that the maximum of the absorbed energy is located in the immediate vicinity of the surface 'illuminated' by radiation. Since the surface of the target is not fixed, free, the area of high pressure inside it cannot exist for long enough. A rarefaction wave is formed on the surface, in which the pressure drops. On the other hand, the zone of increased pressure inside the target generates a compression wave, which then spreads deep into the target.

Consider the process of wave interactions in the acoustic approximation, which is valid for not-too-large temperatures and pressures, when matter does not evaporate, and changes in the speed of sound are not significant. At the initial moment immediately after the instantaneous irradiation, the velocity of the particles of the barrier are equal to zero. In the $p - u$ diagram shown in Fig. 2.11 *a*, their states are described by points lying on the pressure axis. Information on the change of state at each point of the heated layer is propagated by sound disturbances into the target and to its irradiated surface. Subsequent values of the pressure and particle velocity of a substance at each point are found at the intersection of the Riemann trajectories of a change in the state of a substance along the C_+ and C_- characteristics which intersect at a given point at a given time. So, for example, the maximum values of pressure p_∞ and particle velocity u_∞ at a point remote from the surface, where the absorbed energy is zero, correspond on the p, u-diagram to the intersection of the direct $p = \rho c u$ (the initial state for C_- characteristics is $p_0 = 0$, $u = 0$) and $p - pm - \rho c u$ (the initial state for C_+ characteristics $p = p_m$, $u = 0$). Thus, at a distance:

$$u_\infty = \frac{p_m}{\rho c}, \; p_\infty = \rho c u_\infty = \frac{p_m}{2}.$$

The maximum speed of the surface movement towards the incident radiation, as can be seen from Fig. 2.11 *a*, is $u_m = -p_m/\rho c$. As disturbances reach the surface from the inner layers of the barrier, its velocity decreases, as shown in Fig. 2.11 *b*. The expansion of the barrier is accompanied by the appearance inside it of negative pressures, the values of which are at the intersection of the Riemann trajectories for disturbances from the depth of the barrier to its irradiated surface, and disturbances reflected from the surface. In the t–x diagram (Fig. 2.11 *c*), the region of negative

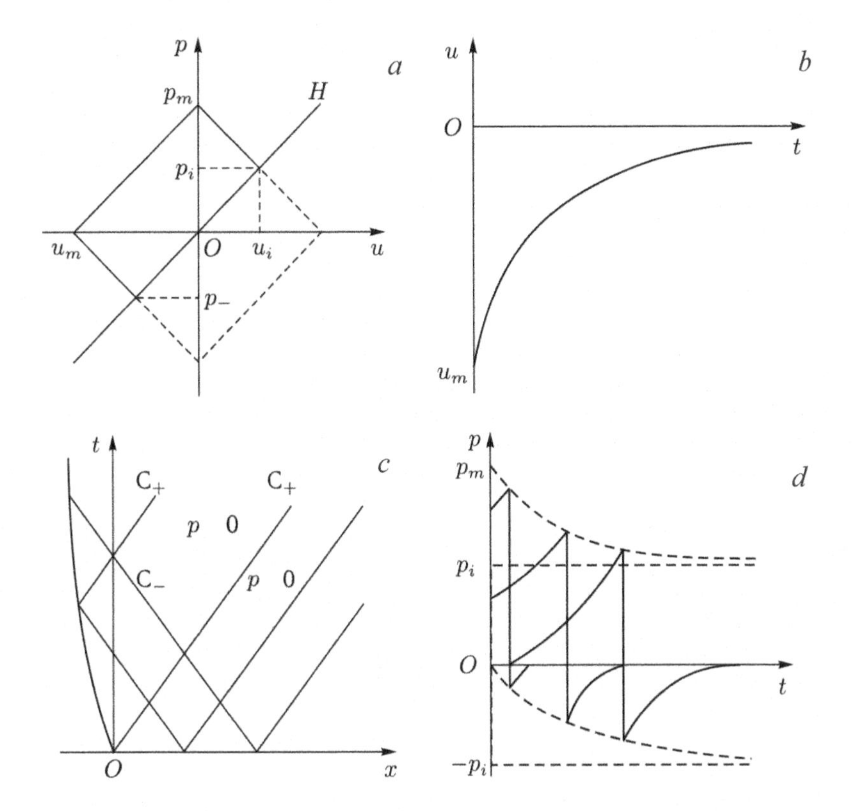

Fig. 2.11. Diagrams of wave interactions and the evolution of the pressure pulse generated by pulsed heat release in the surface layer of the target.

pressures is located above the selected C_+ characteristic, starting from the origin. The maximum of the absolute value of the negative pressure increases with the propagation of the reflected wave into the depth of the obstacle until it reaches the limiting value equal to $p_- = \rho c u_m/2 = -p_m/2$.

The maximum tensile stress is achieved in sections where the absorbed energy and the initial pressure are zero. The change in pressure with time in different sections of the target is shown in Fig. 2.11 *d*. Thus, a fast volume energy release forms an alternating load impulse in the barrier, the amplitude pressure values of which are equal to half the maximum pressure in the energy release zone. This conclusion is confirmed by the results of computer simulation, shown in Fig. 2.12.

From the point of view of studying the mechanical properties of materials, the advantage of laser effects is the possibility of

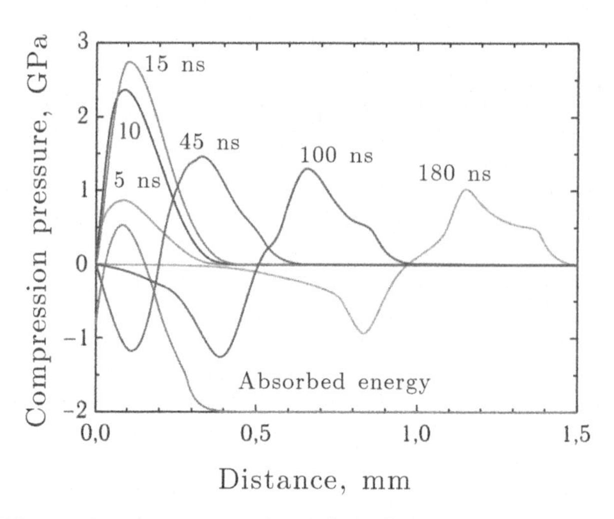

Fig. 2.12. The results of computer simulation of the generation of compression and tension waves in a metal target by a 20-ns electron pulse. Calculation taking into account the elastoplastic properties of the target material.

measuring the resistance to fracture with the shortest durations of a shock load.

The powerful ($\sim 10^{14}$ W) pulsed generators of relativistic electrons and ions created for the purposes of controlled thermonuclear fusion and solving applied problems make it possible to focus intense particle beams in a target with an area diameter of several millimeters. The specific powers introduced in this way of $\sim 10^{14}$–10^{18} W/cm^2 cause rapid heating of the surface layers of the target. The characteristic depth of absorption of electrons with an energy of the order of MeV in metals is 0.1–1 mm. The use of a pulsed generator of ballistically focused proton beams for experiments with shock waves turned out to be fruitful [84].

2.4. Discrete methods for measuring wave and particle velocities

Dynamic diagnostic methods are based on the connection between the quantitative and qualitative parameters of the structure and the evolution of compression and rarefaction waves, which can be fixed experimentally, with the properties of the medium. Measurements of self-similar flows of the type of a steady shock wave or a simple Riemann wave make it possible to determine the properties of the test substance using the kinematic parameters found from experiments, which characterize its response to a shock load.

Historically, the first task for the physics of shock waves in condensed matter was the measurement of the Hugoniots of various substances.

Each experimental point on the Hugoniot is determined from the results of measurements of two independent parameters of shock compression, as a rule, the velocity of the shock wave and the particle velocity behind the shock jump. The pressure, specific volume and specific internal energy of the shock-compressed substance are then calculated on the basis of the laws of conservation of mass, momentum and energy in the form (1.12), (1.13), (1.14). The velocity of the shock wave is determined directly by measuring the time it takes to travel a previously measured distance, usually the sample thickness of the material under study. The overwhelming majority of data on the shock compressibility of condensed media was obtained using the methods of measuring wave and mass velocities based on the use of electrocontact sensors or flare gas gaps.

The self-shorting electric pins record the moments of passage of a shock wave or body surface through the reference points of the U_S or u_p measurement base. When the electric pin is shortened, the simplest electrical circuit produces a current pulse, which is recorded by an electronic oscilloscope. The oscillograms obtained are used to determine the time intervals between the response times of several pins installed along the shock wave path in the sample or on the path of motion of the free surface. The distance between the pin ends is measured with high accuracy, so the measured time intervals easily find the velocity of the shock wave or the speed of movement of the surface of the sample or the impactor. By installing many sensors in one experiment, it is possible to fix the possible tilt and curvature of the wave front, which, after the introduction of appropriate corrections, increases the measurement accuracy. Additional error in the measurement results introduces distortion of signals in the recording equipment and connecting cables.

The method of flare gaps is based on the use of thin (~50 μm) gas gaps between the PMMA (polymethyl methacrylate) block through which the observation is carried out and the surface of the sample. Under the action of a shock wave, the gas in the gap is adiabatically compressed and heated, thus forming a short flash of light. Flashes of gas in the gaps are recorded by a high-speed photo recorder (photochronogap) operating in the streak scan mode. The flash duration is usually 10–50 ns, depending on the thickness of the gap. To obtain a high luminance brightness, argon gap blowing is used, which is especially effective at relatively low pressures. When the intensity of the shock wave in hundreds of kilobars and more fairly good results gives the use of air gaps. The assembly with the sample is made in the form of several steps, so that the flashing gaps are placed at different distances along the shock wave path.

The determination of the particle velocity is usually based on the analysis of the decomposition of a discontinuiy at the boundary between the impactor and the target, or between a base plate of reference material and a sample. If the impactor and the sample are made of the same material, then due to the symmetry the magnitude of the particle velocity is exactly equal to half the impact velocity. In other cases, the 'braking'method or the 'reflection' method are used to determine the particle velocity. The braking method is used to construct the Hugoniot from the measured values of the impactor velocity u_i and the shock wave velocity U_S in the sample–target. The Hugoniot of the impactor material is assumed to be known from the independent prior measurements. As a result of a collision shock waves are formed in the impactor and the target and diverge from the contact surface between these objects. The pressure and particle velocity on both sides of the contact surface are equal and satisfy the Rankin – Hugoniot conservation laws and the Hugoniots of both the impactor and the target. Their values are sought at the intersection in the coordinates p–u_p of the Hugoniot of braking the material of the impactor and the wave beam $p = \rho_0 U_S u_p$ for the shock wave in the target (Fig. 2.13).

The reflection method analyzes the reflection of the shock wave in the reference base plate from the contact surface with the sample. The initial parameters of the shock wave in the base plate and the equation of state of the latter are assumed to be known from independent measurements. As a result of reflection two waves are formed on the contact surface – a shock wave in the sample and, depending on the dynamic impedance ratio, a reflected shock wave or a rarefaction wave in the base plate. For shock waves of moderate intensity with short-range extrapolations of the state of a substance in the reflected compression and rarefaction waves they are accurately described in the coordinates of the p–u_p by a curve, which is mirror-symmetric to the Hugoniot. According to the measured velocity of the shock wave in the sample U_S and the known parameters of the shock wave in the base plate, the pressure and particle velocity on

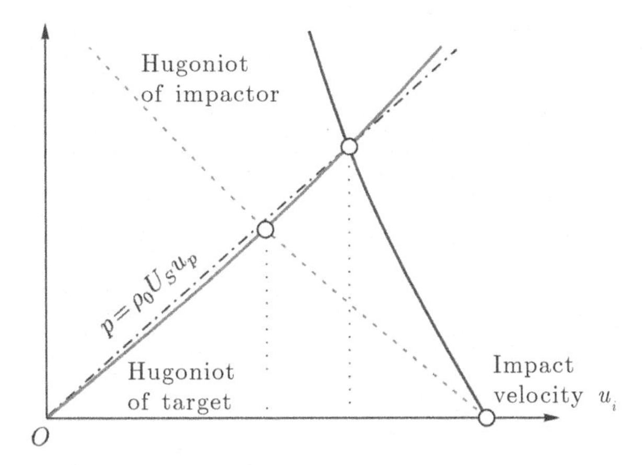

Fig. 2.13. Measurement of Hugoniot by the braking method.

the contact surface are found at the intersection of the wave beam $p = \rho_0 U_S u_p$ with specular reflection of the Hugoniot of the base plate substance.

Thus, the braking method and the reflection method are relative methods, the accuracy of which directly depends on the error of the equation of state of the reference material of the impactor or base plate. Experiments with the collision of plates of the same material in an identical initial state provide absolute measurements of the particle velocity and other parameters of shock compression.

In a similar way, the isentropes of unloading a substance from a shock-compressed state are measured. To do this, low impedance barriers of substances, including gaseous ones, whose Hugoniots are known with good accuracy, are located directly behind the sample. Measurements of the velocity of shock waves in the barriers give in fact the values of pressure and particle velocity at the Riemann unloading isentrope (Fig. 2.14). A set of such data is then used to check and refine the equation of state of the test substance.

Along with discrete measurements of the kinematic parameters of shock waves in high-dynamic pressure physics, continuous recording of wave profiles of pressure and particle velocity of a substance is widely used. These measurements are used to study the elastic-plastic and strength properties of condensed media, the parameters of phase transitions and chemical transformations in shock waves.

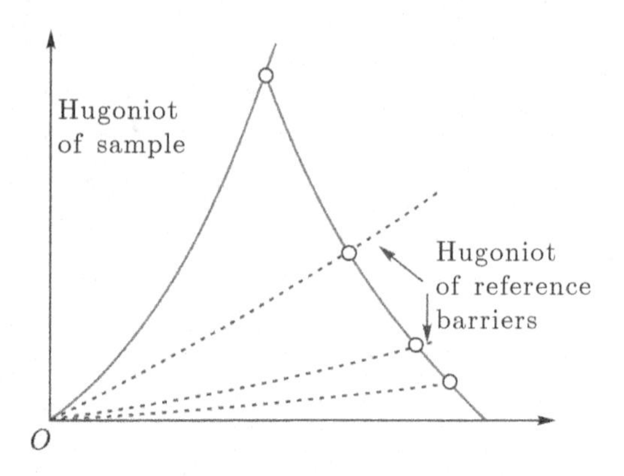

Fig. 2.14. Measurement of isentropic unloading of shock-compressed matter by the method of reference barriers.

2.5. Methods for continuous recording of particle velocity histories

The monitoring of particle velocity histories in dielectric materials is carried out by the magnetoelectric method. The method was proposed and developed by Academician E.K. Zavoisky initially to determine the parameters of detonation of explosives used in the creation of the atomic bomb. For measuring the particle velocity history in the sample of the test substance, a U-shaped gauge made of thin aluminium foil is installed. The experimental assembly is placed in a uniform magnetic field so that the gauge 'crossbar', which is its sensitive element, is perpendicular to the direction of the field lines and parallel to the shock wave front (Fig. 2.15).

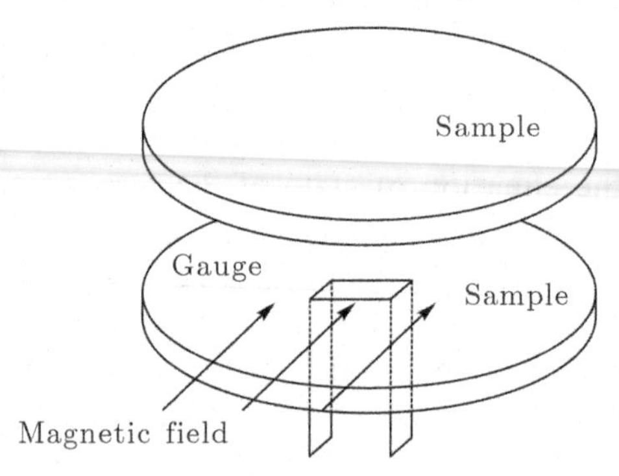

Fig. 2.15. Scheme of measurements of particle velocity histories by the electromagnetic method.

In the load pulse, the gauge is carried away in motion along with the substance surrounding it. At the same time the EMF of magnetic induction is induced at its terminals equal to:

$$U(t) = -u(t)\,Bl,$$

where l is the length of the sensor 'crossbar' (usually ~10 mm), B is the magnetic field induction (~1 Tesla). The magnetoelectric method for recording particle velocity histories is widely used in experiments with explosives. Both stationary and pulsed disposable solenoids are used as sources of the magnetic field. Methods have been developed for simultaneously recording up to ten to twelve particle velocity histories, and thus, in one experiment, to measure the evolution of a shock compression pulse.

2.5.1. Capacitor gauge of surface velocity

The method of capacitive sensors [5] is designed to measure the velocity histories of the surface of metal samples. An example of experiments with the use of a capacitor gage is shown in Fig. 2.16. A flat electrode is installed at a distance x_0 from the surface of the sample and forms with the sample the measuring capacitor C with a capacitance of several picofarads. The voltage E is supplied to the capacitor from the EMF source through the load resistance R_{in}, the value of which (50 Ohms) is chosen small enough so that the time

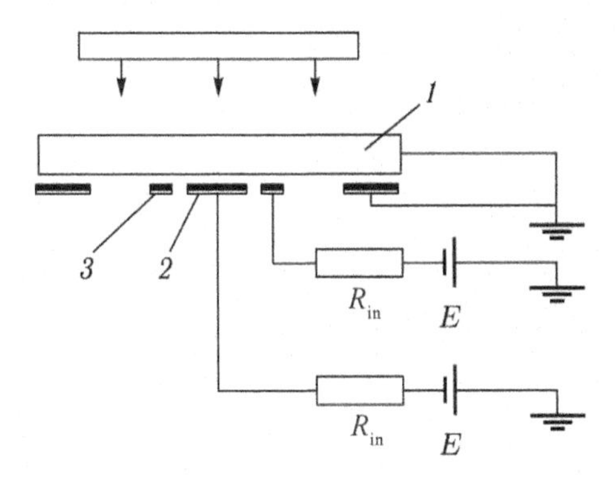

Fig. 2.16. The scheme of experiments on the recording of the velocity of free surface of samples by the method of capacitor gauge: 1 – loaded sample, 2 – measuring electrode, 3 – protective ring of capacitor gauge, E – the source of EMF, R – load resistance.

constant of the circuit is much less than the characteristic time of recording. A guard ring is used to maintain the uniformity of the electric field in the area of the measuring electrode.

When the shock wave moves to the surface of the sample after the shock wave arrives, the capacitance of the measuring capacitor begins to change, and a charging current appears in the gauge circuit, which is recorded during the measurement. Its strength is proportional to the rate of change of capacitance or the velocity of movement of the surface of the sample u_{fs}:

$$i(t) = E\frac{dC}{dt} = \frac{\varepsilon SE}{4\pi x^2(t)dt}\frac{dx}{dt} = \frac{\varepsilon SE}{4\pi x^2(t)}u_{fs}(t), \quad (2.2)$$

where E is the voltage at the EMF source, ε is the dielectric constant, S is the area of the measuring electrode, x is the distance between the plates at time t, calculated by integrating the velocity during the processing of the experimental oscillogram $i(t)$.

Capacitors with a capacitance of about 0.01 microfarads, charged to a voltage of 3 kV, are used as EMF sources. With such a voltage at the source, the characteristic signal level of a capacitive sensor is units up to tens of millivolts. Figure 2.17 shows a typical current waveform, obtained in the experiment with a capacitor gauge, and the result of its processing.

This method is contactless, so its resolution is limited, in principle, only by the non-simultaneity of the output of the recorded load pulse in the area of the sample surface monitored by the gauge.

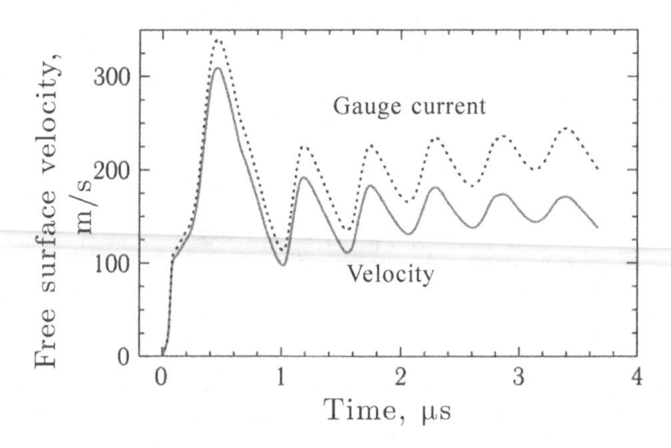

Fig. 2.17. An example of an oscillogram of an experiment with a capacitor gauge sensor and the free surface velocity history of a 35Kh3NM steel sample obtained from its processing when impacted by a thin aluminium plate.

Depending on the required resolution and the total recording time, the diameter of the measuring electrode and the distance between it and the sample surface vary between 5–25 mm and 1–6 mm, respectively. With a minimum diameter of the measuring electrode of 5 mm, the actual resolution of the gauge was 10–15 ns in experiments. This temporal resolution for a capacitor gauge seems to be the ultimate. Further reduction of the diameter of the measuring electrode is associated with increased uncertainty in the interpretation of measurement results.

2.5.2. Laser measuring devices for the velocity of motion of free and contact surfaces of samples

The use of lasers for measuring the velocity of a substance in experiments with shock waves is based on the use of the Doppler effect. Since at a speed of movement of the reflecting surface of ~100–1000 m/s the effect is very small (the wavelength shift is ~10^{-2}–10^{-1} Å), two-beam or multipke-beam interferometers are used to fix it. In this case, the measurements take on a differential character, which significantly increases their accuracy. The high spatial resolution of the laser methods is ensured by the fact that the probing laser radiation is focused on the sample under investigation into a spot with a diameter of 0.1 mm.

Figure 2.18 is a scheme of a laser Doppler velocity meter VISAR (Velocity Interferometric System for Any Reflection) [6]. The

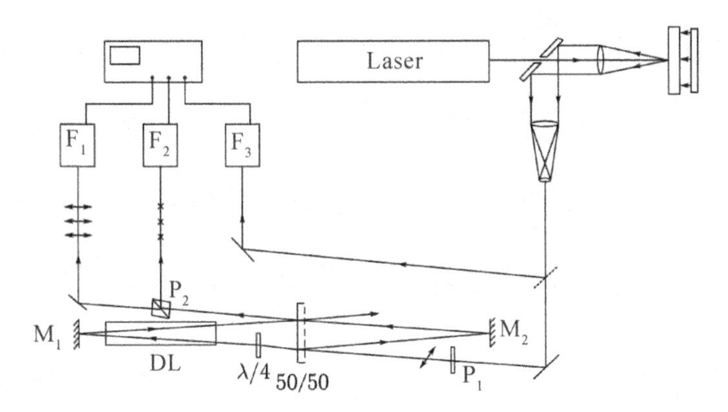

Fig. 2.18. The scheme of recording of wave profiles with a laser Doppler velocity meter. M_1, M_2 – 100% end mirrors; 50/50 – 50% beam splitter; DL – glass delay line; P_1 – polarizer oriented at an angle of 45° to the horizontal; P_2 – polarizing beam splitter; λ/4 – quarter-wave plate; $F_{1,2,3}$ – receivers of output radiation and control of intensity.

measurement of the velocity of the reflecting surface is carried out by recording periodic changes in intensity (fringes) at the interference of two beams of coherent light with close wavelengths. In this case, the rays of light reflected from the moving surface at different moments in time interfere. If the speed of the reflecting surface changes, then due to a shift in time, the magnitude of the Doppler effect for interfering beams turns out to be different. The intensity fringes recorded by the photoreceivers have a frequency proportional to the acceleration of the reflecting surface and the value of the relative time shift.

The idea of the method is implemented in the instrument as follows. The single-frequency laser radiation is focused on the sample surface. The light reflected from the controlled surface is collected by the lens into a quasi-parallel beam and directed to the interferometer. After narrowing in the input telescope, the beam of reflected light is split by the beam splitter into two beams of equal intensity, which are sent to the optically symmetric 'legs' of the interferometer. One of the legs contains a glass delay line, as a result of which the time of the double passage of light in this leg is longer than in the opposite one, for some delay Δt. After reflection from the end mirrors, both beams return to the beam splitter where their interference occurs.

To meet the requirement of parallelism of the wave fronts of recombining beams, the geometric paths of light in the two legs of the interferometer must differ by a strictly defined value $\Delta L = l(1-1/n)$, where l, n is the length and refractive index of the delay line. In this case, the delay time is:

$$\Delta t = \frac{2l}{c}(n - 1/n), \tag{2.3}$$

where c is the speed of light in a vacuum. Due to the optical symmetry of this interferometer, the spatial coherence of the analyzed radiation is not required. A good interference contrast is obtained even when working with light reflected from a scattering, diffuse surface, when the beam entering the interferometer has a speckled transverse structure.

The current value of the monotonically varying speed of the reflecting surface $u(t)$ is determined by the number of fringes of light intensity $N(t)$ registered by photoreceivers from the relation

$$u(t) = \frac{\lambda}{2\Delta t}\frac{N(t)}{1+\delta}, \quad \delta = \frac{n}{n^2-1}\lambda\frac{dn}{d\lambda}, \tag{2.4}$$

where λ is the wavelength of the probe radiation, n is the refractive index of the glass of the delay line, the number of fringes N can be a non-integer number. The magnitude of the velocity increment per fringe $\Delta u = \lambda/(2\Delta t(1 + \delta))$ is called the velocity-per-fringe (VPF) constant, the value of which varies from tens to thousands of meters per second.

For reliable fixation of changes in the sign of acceleration with a non-monotonic change of $u(t)$, a polarization coding system is provided in the device. Before splitting in a large beam splitter, the beam of light reflected from the sample is polarized at an angle of 45° to the horizontal. One of the interferometer legs contains a quarter-wave plate, in which the phase of vertically polarized light is shifted by 90° relative to the horizontal component. After the beams recombine at the output of the interferometer, the light beam is split by the polarization beam splitter into two components with vertical and horizontal polarization. Two photoreceivers independently record the ocillations of the intensity of each component, which are also shifted in phase relative to each other by 90°. As a result, a change in the sign of acceleration will inevitably be fixed by at least one photoreceiver by changing (outside the extremes of the fringes) the sign of the change in the luminous flux, and the direction of the change in velocity is easily determined from the ratio of the phases of the interference fringes.

The change in the light flux at the output of the interferometer is associated with a change in the velocity of the reflecting surface by a sinusoidal dependence. Based on this, the current values of velocity can be determined from experimental waveforms not only discretely – by counting the number of fringes, but also by measuring the instantaneous values of the relative light fluxes in each recording channel within individual fringes. Thanks to the controllability of the light entering the device and polarization coding, the error in speed measurements exceeding the $\lambda/2\Delta t$ value can be reduced to 1% or less. The best time resolution achieved when measuring the width of the shock wave front in metals by this method was approximately 1 ns.

Laser interferometric measurements of the velocity of a substance in shock waves are used both for recording the speed of movement of the free surface of the body, and for measuring wave profiles inside a transparent medium or on the contact surface between the test sample and the 'window' of transparent material. In this case, one should take into account the effect of shock compression of a transparent medium on its optical characteristics and patterns of light reflections in it from a moving surface.

In the case of working with windows, the probing radiation passes through the surface of the window, then through the moving shock front, behind which the substance has a modified refractive index, after which it is reflected inside the shock-compressed substance and goes out through the shock front and the stationary surface. The transition of each boundary is accompanied by a change in the speed of light and its wavelength. In the simplest approximation, it can be assumed that the difference in the length of the optical path in a substance and in a vacuum with the same geometric length of the path is proportional to the integral density of the substance within this length of the path, expressed by the number of atoms per 1 cm^3, regardless of the compression of the substance. This approximation is called the Gladstone–Dale model. In this case, the value of the Doppler shift of the radiation wavelength when it is reflected in a shock-compressed transparent medium exactly corresponds to that which occurs at the same velocity in a vacuum. If the Gladstone–Dale model for the selected window material is not accurate enough, then appropriate corrections should be taken into account in the processing of interferograms. As a result, the main calculation formula for a two-beam interferometer takes the form

$$u(t) = \frac{\lambda}{2\Delta t} \frac{N(t)}{(1+\delta)(1+\Delta v / v_0)}, \tag{2.5}$$

where $\Delta v/v_0$ is the frequency correction associated with the effect of compression on the refractive index.

The speed of the reflecting surface $u(t)$ is obtained as a result of joint processing of three oscillograms. In the linear operation of photomultipliers their readings $i(t)$ are related to the current value of the measured velocity $u(t)$, the relative intensity of the light entering the interferometer $A(t)$ and the value of the interference contrast $K(t)$ by the relations:

$$\begin{aligned}
i_1(t) &= \frac{1}{2} J_{10} A(t) \left\{ 1 + K(t) \sin\left[2\pi \frac{u(t)}{\Delta u} + \varphi \right] \right\} + i_1^{\min} A(t), \\
i_2(t) &= \frac{1}{2} J_{20} A(t) \left\{ 1 + K(t) \sin\left[2\pi \frac{u(t)}{\Delta u} + \varphi + \theta \right] \right\} + i_2^{\min} A(t), \\
i_3(t) &= \frac{1}{2} J_{30} A(t).
\end{aligned} \tag{2.6}$$

In these ratios Δu is the VPF interferometer constant, φ is the initial beat phase, θ is the phase shift between the intensity fringes of the vertically and horizontally polarized light components at the interferometer output, J_{10}, J_{20} is the initial fringe amplitude, J_{30} is the initial intensity of the light entering the interferometer, i_1^{min}, i_2^{min} are deviations of oscillograph beams in the fringe minima, the magnitude of which, due to the imperfection of the optical coatings and the errors of the electronic part of the device, is usually $i^{min} \approx 0.1J_0$. The magnitude of the interference contrast $K(t)$ can decrease from 1 to 0.5–0.7 due to the displacement of spots in the interfering beams and the inhomogeneity of the surface movement within the probe beam. The values of θ and φ are determined from the oscillograms by the values of i_1 and i_2 before the beginning of the movement of the sample surface taking into account the direction of their subsequent change.

As an illustration Fig. 2.19 shows experimental oscillograms of the interferometric recording of a compression pulse in a sample of VT6 titanium with a thickness of 10 mm when an aluminium plate with a thickness of 2 mm collides with it at a speed of ~700 m/s and the velocity history of the free surface obtained from their processing. The VPF constant speed of the interferometer in this experiment was 305 m/s.

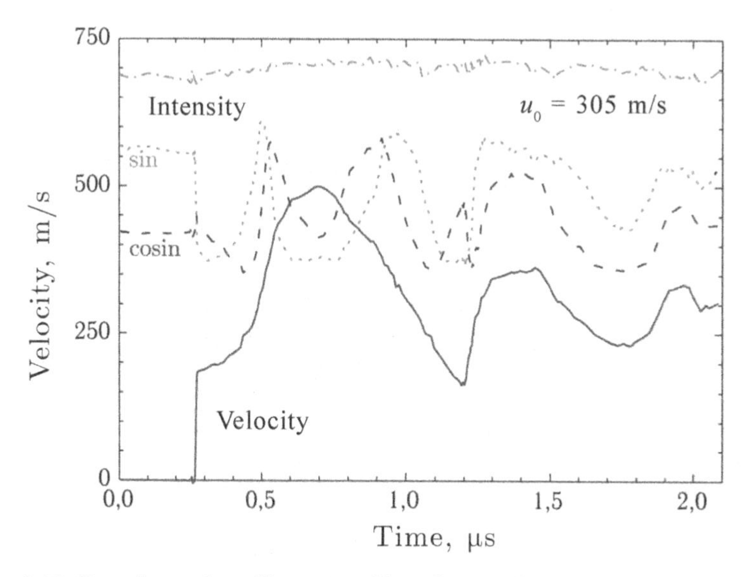

Fig. 2.19. Experimental oscillograms of interferometric recording of the free surface velocity history of a titanium alloy plate and the result of their joint processing.

In experiments with strong shock waves, the rise time of the velocity in a shock wave may be less than the delay time in VISAR and go beyond the time resolution of the recording equipment. In this case, it becomes impossible to determine directly from the oscillograms the absolute values of the velocity behind the shock wave. In this case, to determine the number of 'missed' fringes, analysis of the decomposition of the discontinuity, numerical simulation of the experimental situation, and other additional considerations are involved.

2.5.3. Manganin piezoresistors for recording pressure histories of shock compression

The main method of recording mechanical stress or pressure is currently the method of manganin gauges. The use of manganin gauges is based on the high sensitivity of the resistivity of manganin to pressure, and on low sensitivity to temperature changes. Usually, the gauge is made in the form of a flat zigzag tape with a thickness of 10–30 μm, occupying an area of ~0.1–1 cm^2. The sample is made multipart; the gauge is laid between the sample plates and, if necessary, is separated from them by insulating films (Fig. 2.20). During the measurements, a current is passed through the gauge, a voltage drop across it is recorded with an oscilloscope, which increases with increasing pressure acting on the gauge. To improve the signal-to-electrical noise ratio and eliminate gauge overheating, pulsed current sources of 5–10 A and duration of ~100 μs are used.

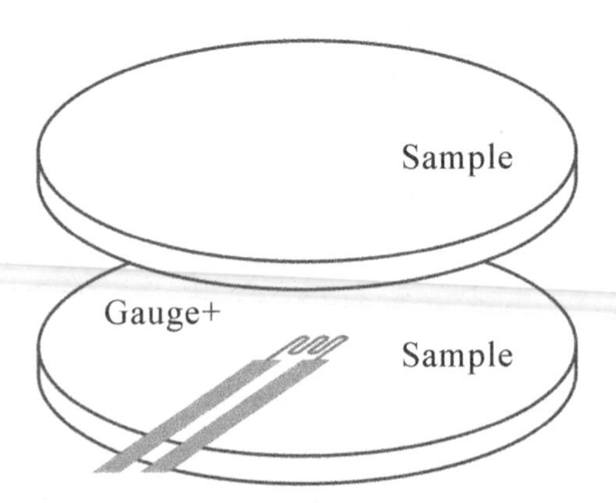

Fig. 2.20. The location of the manganin pressure gauge in the sample.

The gauge is included in the resistance bridge or other differential registration schemes are used in order to eliminate the constant component of the signal, determined by the initial resistance of the gauge, and thereby improve the measurement accuracy. The two-point scheme is used for relatively high-resistance (5–50 ohms) gauges. In some cases, it is advisable to use gauges with an initial resistance at the level of tenths or hundredths of Ohms. Such gauges have, in particular, the advantage that their readings are less sensitive to the shunting effect of the electrical conductivity of the environment. Low-resistance gauges are included in the measuring circuit in a four-point scheme.

Due to the destructive action of the shock waves, it is impossible to calibrate each gauge used. For this reason, to determine the pressure, a single dependence of the relative change in the electrical resistance $\Delta R/R_0$ on the shock compression pressure is used, common to all gauges from manganin of this brand. The calibration dependence is based on the results of experiments with the placement of gauges in reference materials with the well-known compressibility. In the assemblies, shock waves are excited and their kinematic parameters are determined in independent ways. Special measurements have shown that in the pressure range not lower than 7–10 GPa the change in the resistance of manganin is practically reversible and does not depend on whether the dynamic compression is shock, stepwise or isentropic. Unloading to zero pressure is associated with a small hysteresis of the resistance of the manganin gauges. The irreversible component of the manganin resistivity increment is associated with the material hardening under shock-wave compression and does not exceed 2.5% of the initial resistance. Annealing of the manganin leads to an increase in the amplitude values and hysteresis of the gauge readings by the same value.

Figure 2.21 shows the dependence of electrical resistance on pressure for manganin of the MNMtsAZh 3-12-0.25-0.2 brand. Subtracting from the experimental dependence of the irreversible component the change in resistance yields a curve close to the measurement results under hydrostatic compression (Fig. 2.11, b).

Since, for carrying out measurements, gaskets of insulating material are introduced into the sample along with the gauges, the indications of the gauges may have some inertia. The distortions are determined by the settling time (in the process of multiple reflections of the waves in the gaskets) of the pressure in the insulation,

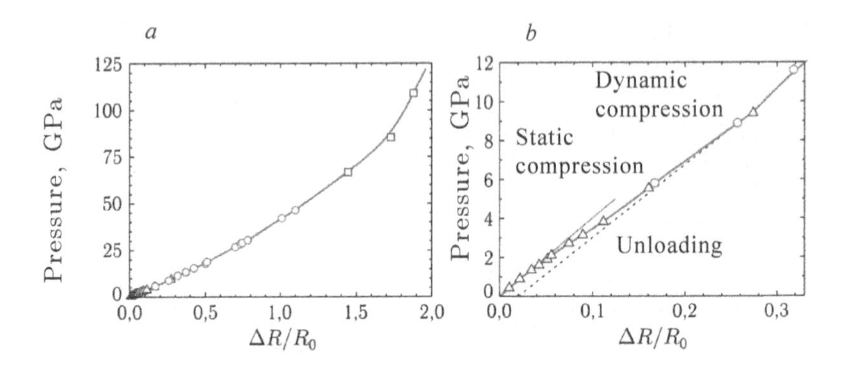

Fig. 2.21. Dependence the relative increment of electrical resistance of manganin R/R0 on the impact compression pressure. *a* – general view in a wide pressure range. Different markers shos data by different authors. *b* – initial section of the dependence. Hysteresis and piezocoefficient in static compression are also shown.

equal to the compressive stress in the environment in the direction perpendicular to the gauge plane. The inertia is especially significant when recording waves of low intensity. With increasing pressure, the distortion of the recorded waveform decreases due to an increase in the speed of sound in insulating gaskets. When registering wave profiles simultaneously in two or more sections of the sample, the distortions accumulate from the gauge to the gauge.

As an illustration of the use of manganin pressure gauges, Fig. 2.22 presents the results of measurements of wave profiles in iron and steel under shock compression to pressures below and above their transformations into the high pressure phase. An increase in compressibility during transformation leads to splitting of the shock wave with the formation of a three-wave configuration and, more interestingly, a rarefaction shock wave during unloading. The rarefaction shock wave is a rather exotic phenomenon, which prior to these measurements was identified by indirect manifestations. Direct registration of the shock rarefaction wave is not only illustrative, but also makes it possible to determine the parameters of the reverse transformation into the low pressure phase.

Under conditions of non-one-dimensional loading, a change in the length of the gauge's sensitive element leads to the appearance of an additional increment of electrical resistance. For the separation of pressure and strain measurements are carried out using geometrically identical gauges made of materials with significantly different dependences of electrical resistivity on pressure. Then, having two

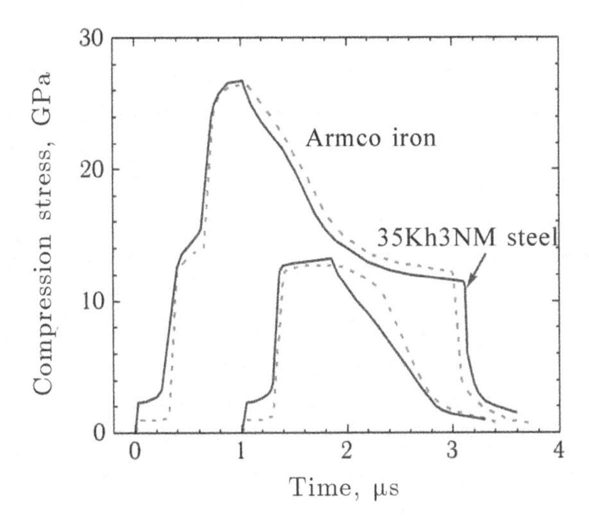

Fig. 2.22. Pressure histories in iron and steel samples under shock compression below and above the transformation pressure $\alpha \rightarrow \varepsilon$.

oscillograms and two unknown quantities – the pressure and the length of the sensitive element, it is easy to calculate the real pressure history $p(t)$. To this end, in addition to manganin, gaugesfrom Constantan are used; the piezoelectric resistivity coefficient of this material is significantly lower than that of manganin.

The described method was used, in particular, for measuring pressure histories in cylindrical inserts located on the axis of the detonating explosive charge. Under these conditions, a detonation wave, sliding along the generator of the cylinder, excites in it a conical converging shock wave. The reflection of a conical shock wave on the cylinder axis is irregular and is accompanied by the formation of a concave Mach disk. At some distance, the process is stabilized – in a cylindrical insert, a stationary shock-wave configuration is formed, having a shape close to a truncated cone and propagating at an explosive charge detonation velocity. Figure 2.23 presents the results of measurements of pressure histories with irregular reflection of a conical shock wave in cylindrical plexiglass inserts placed on the axis of RDX charges. Sensitive elements of the gauges were made in the form of open rings and were placed in the sample coaxially in order to ensure synchronism of loading. The measurement results showed that, in contrast to the classical mode of irregular reflection of strong shock waves, the wave configuration in this case does not contain a reflected shock wave.

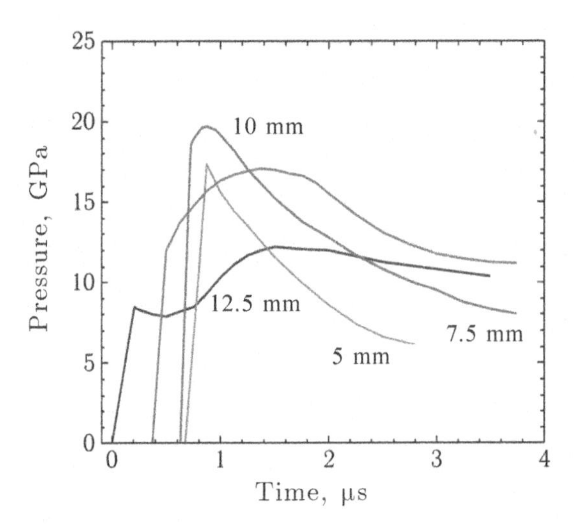

Fig. 2.23. Pressure histories with irregular reflection of a conical shock wave in cylindrical plexiglass inserts placed on the axis of hexogen charges. The values of the radii of the sensors are shown.

Manganin pressure gauges are widely used in experiments with shock waves. With their help, measurements are carried out at low (up to 77 K) and elevated (up to 600 K) initial temperatures of the samples. The elastic–plastic properties and polymorphic transformations of substances, the evolution of compression pulses in reacting explosive materials are studied, the parameters of dynamic load are determined under technological conditions of explosive processing. It should be noted that the manganin is not the only piezoresistive material used as pressure gauges. At low pressures (up to 1.5–2 GPa), ytterbium gauges are used, the piezo-sensitivity of which is much higher than that of manganin. The scope of ytterbium is limited to polymorphic transformation at a pressure of about 4 GPa.

The manganin piezoelectric resistors in various configurations are the main tool for measuring the pressure profiles of shock compression of solids. In addition, there are quartz and polymer sensors, whose work is based on the effect of piezoelectricity, film sensors calibrated to change their capacitance during compression, but they are used much less widely than piezoresistors.

3

Main directions of investigations of the behaviour of condensed matter under shock compression using the methods of continuum mechanics

3.1. The structure of compression and rarefaction waves in an elastoplastic body

The equations of the continuous one-dimensional motion of an elastoplastic medium are obtained from the equations of gas dynamics by replacing pressure with a normal stress [7]:

$$\rho_0 \frac{\partial V}{\partial t} - \frac{\partial u_p}{\partial h} = 0, \; \rho_0 \frac{\partial u_p}{\partial t} + \frac{\partial \sigma_x}{\partial h} = 0, \; \frac{\partial E}{\partial t} = -\sigma_x \frac{\partial V}{\partial t}, \tag{3.1}$$

where ρ_0 is the density of the medium at zero pressure, $V = 1/\rho$ is the specific volume, u_p is the particle velocity of the substance, σ_x is the normal stress acting in the axial direction, E is the specific internal energy, t is the time, h is the Lagrange (substantial) coordinate. When analyzing shock-wave phenomena in solids, stresses are usually taken as positive in compression, as in geophysics.

In the range of moderately high stresses, the shock compression of solids has a substantially elastoplastic character. A variety of flow criteria are used to determine the limits of elastic deformation. The task of the latter is to establish, on the basis of simple standard tests, the conditions under which plastic deformation begins in the material. In particular, according to the Coulomb and Gest hypotheses, the the ultimate elastic state at a given point of the continuum occurs when the greatest shear stress τ reaches the value corresponding to the ultimate elastic state of the same material with simple tension:

$$\left|\tau \leq \frac{\sigma_Y}{2}\right|. \tag{3.2}$$

where σ_Y is the yield strength of the material. According to the Huber and von Mises hypotheses, the ultimate elastic state at the point of the continuous medium occurs when the specific energy of the shape changes reaches the value corresponding to this energy at simple tension:

$$\frac{1}{2}\left[\left(\sigma_x-\sigma_y\right)^2+\left(\sigma_y-\sigma_z\right)^2+\left(\sigma_z-\sigma_x\right)^2+6\left(\tau_{xy}^2+\tau_{yz}^2+\tau_{zx}^2\right)\right] \leq \sigma_Y^2. \tag{3.3}$$

In the simplest cases of uniaxial stress and uniaxial strain, these two hypotheses give identical results.

In the plastic region, the strain increment along each axis is equal to the sum of the elastic and plastic components:

$$d\varepsilon_k = d\varepsilon_k^{el} + d\varepsilon_k^{pl}. \tag{3.4}$$

Plastic strains are not accompanied by a change in volume:

$$d\dot{\varepsilon}_x^{pl} + d\dot{\varepsilon}_y^{pl} + d\dot{\varepsilon}_z^{pl} = 0. \tag{3.5}$$

The increment of the total maximum shear strain γ is expressed as

$$d\gamma = d\varepsilon_x - d\varepsilon_y = d\tau / G + d\gamma_p. \tag{3.6}$$

where γ_p is the plastic component of shear strain.

Under the conditions of standard tests under the uniaxially stressed state, the idealized deformation diagram of an elastoplastic body has the form shown in Fig. 3.1 *a*. In this case:

$$-\sigma_Y \leq \sigma_x \leq \sigma_Y;\ \sigma_y = \sigma_z = 0;\ \varepsilon_y = \varepsilon_x \neq 0 \tag{3.7}$$

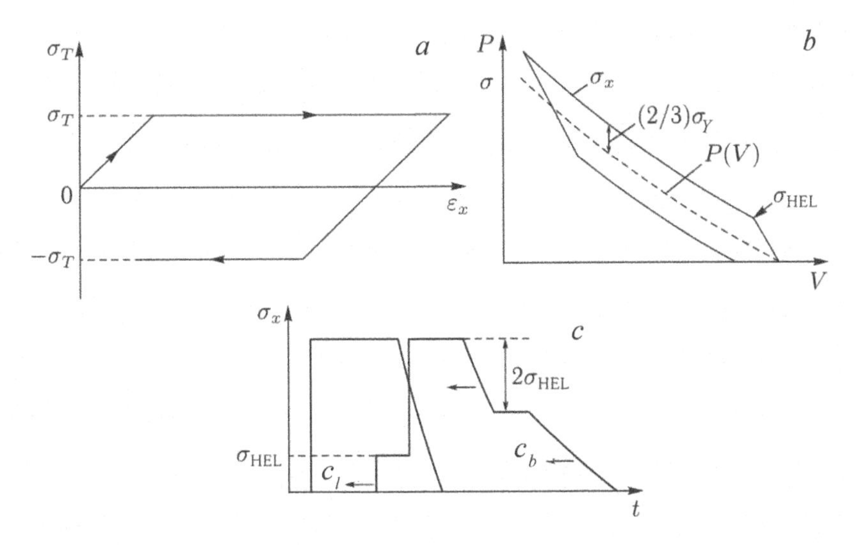

Fig. 3.1. Features of elastoplastic deformation of solids under conditions of uniaxially stressed state and under conditions of uniaxial deformation.

The deformation has an elastic character until the stress reaches the value of the yield strength σ_Y. In this domain, the response of the material to the load is described by Young's law. In the field of plastic deformation, the stress remains unchanged: $\sigma_x = \sigma_Y$. When changing the direction of deformation to the opposite the material again behaves elastically up to the fulfillment of the yield condition in the stress region of the opposite sign.

In plane compression and rarefaction waves, the loading conditions are characterized by one-dimensional deformation: $\varepsilon_y = \varepsilon_z = 0$; in this case, the normal stresses in the directions perpendicular to the direction of compression are not zero: $\sigma_y = \sigma_z \neq 0$. The change in the stress state in the compression–rarefaction cycle is shown in Fig. 3.1 *b* for this case. In the elastic domain, the longitudinal compressibility of the material is equal to

$$-\frac{1}{V}\frac{dV}{d\sigma_x} = \frac{1}{K + 4G/3}, \tag{3.8}$$

less bulk compressibility

$$-\frac{1}{V}\frac{dV}{dp} = \frac{1}{K}, \tag{3.9}$$

Here K is the bulk modulus, G is the shear modulus, $p = (\sigma_x + \sigma_y + \sigma_z)/3$ is the pressure (the spherical part of the stress tensor). The yield condition is satisfied at $|\tau| = 1/2\sigma_Y$, i.e.

$$|\sigma_x - p| = 2/3\sigma_Y. \quad (3.10)$$

The longitudinal compressibility in the plastic domain is equal to the bulk.

The transition from elastic to plastic deformation occurs when stress is reached

$$\sigma_x = \sigma_{HEL} = \sigma_Y(K/2G + 2/3), \quad (3.11)$$

where σ_{HEL} is the elastic limit under uniaxial shock compression (Hugoniot Elastic Limit, HEL). During unloading, the elastic deformation section is twice as large as $2\sigma_{HEL}$, since the shear stresses decrease to zero in the rarefaction wave, then the sign changes at τ and the absolute value of τ increases to the limit value $|\tau_T| = \sigma_Y/2$.

Figure 3.1 *c* shows the evolution of the originally rectangular compression pulse in an idealized elastoplastic material. Due to the difference in the longitudinal compressibility in the elastic and plastic regions of deformation, the compression and rarefaction waves split with the formation of elastic precursors propagating with the longitudinal sound speed: $c_l = \sqrt{(K + 4G/3)/\rho}$. The compressive stress behind the front of the elastic precursor is determined by the yield strength of the material and is equal to σ_{HEL}. The minimum propagation velocity of longitudinal perturbations in the plastic region is determined by the longitudinal sound speed: $c_b = \sqrt{K/\rho}$. The compressive stress and other state parameters behind the plastic shock wave are determined by the conditions of shock-wave loading, in particular, by the impact velocity. Due to the nonlinear compressibility of materials, the velocities of elastic and plastic shock waves usually exceed the values of the longitudinal and bulk speeds of sound in the initial state. The velocity of the plastic shock wave U_S increases with increasing pressure and, at $U_S > c_l$, the two-wave compression configuration disappears.

Using the results of measurements of the free surface velocity history, the longitudinal stress at the front of the elastic precursor or the Hugoniot elastic limit is defined as

$$\sigma_{\text{HEL}} = 0.5u_{\text{HEL}}\rho_0 c_l, \tag{3.12}$$

where u_{HEL} is the free surface velocity jump in the precursor front.

The elastic limit for one-dimensional deformation is associated with the yield strength in the usual sense of σ_Y by

$$\sigma_Y = \frac{3}{2}\sigma_{\text{HEL}}\left(1 - c_b^2 / c_l^2\right). \tag{3.13}$$

3.2. Compression wave in hardening and softening materials

In addition to the dynamic elastic limit, a detailed analysis of the structure of an elastoplastic compression wave can provide information on the work hardening and relaxation properties of a material under high strain rate conditions. Strain hardening (by increasing the flow stress τ in the plastic region) manifests itself in the shape of an elastic precursor profile. The viscosity or relaxation of stresses causes some decay of the elastic precursor as it propagates and reduces the steepness of the plastic shock wave.

To clarify the connection between the precursor form and the material deformation diagram, Fig. 3.2 shows idealized stress–strain diagrams corresponding to cases of ideal plasticity, strain hardening, and temporary softening with the formation of upper and lower yield strengths.

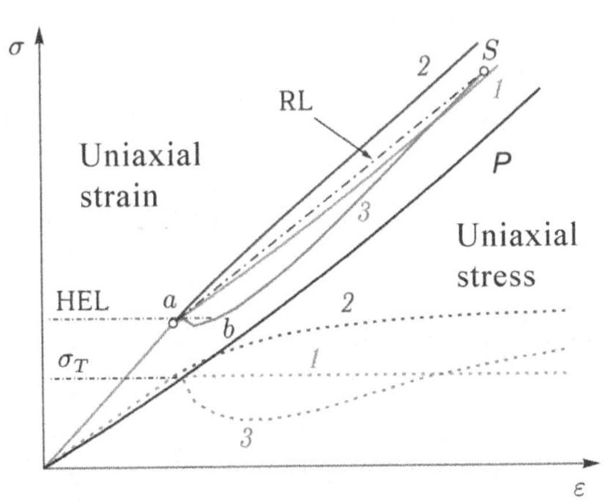

Fig. 3.2. Diagrams of elastoplastic compression of materials with perfect plasticity (to the left 1), strain hardening (curves 2) and initial softening (curves 3). The dashed lines correspond to the conditions of a uniaxially stressed state, the solid lines correspond to uniaxial strains. Beam *aS* shows an example of a Rayleigh line (RL).

For an ideal elastoplastic material and a softening material (curves 1 and 3 in Fig. 3.2), the initial state for the Rayleigh line of the second compression wave coincides with the state at the Hugoniot elastic limit. It is important to note that the Rayleigh line in these coordinates can have only a positive or zero slope and cannot have intermediate intersections with the deformation diagram. For this reason, in particular, the states on the stress–strain curve of the softening material that are below the horizontal line *ab* cannot be realized in a compression wave. As a result, the so-called 'sharp yield point, which is observed in the diagrams of low-rate deformation of iron and some low-carbon steels, cannot by itself lead to the formation of a similar feature in the structure of an elastoplastic compression wave. However, the stress spike at the front of the elastic precursor in experiments is recorded quite often. Its formation is associated with specific features of stress relaxation, discussed in the next section. In this case, as an indicator of temporary softening, there may be a small, less than the bulk sound velocity c_b, the propagation velocity of the plastic compression wave.

In analyzing the nature of the flow in the region between the elastic precursor and the plastic compression wave, we take into account that abrupt turn in the compressibility of a substance correspond to discontinuities in the dependence of sound velocity on stress. In turn, a jump in the speed of sound leads to the formation of flow sections with constant parameters. In other words, if the onset of plastic deformation is associated with a sharp turn of the stress–strain diagram, then the elastic precursor of the compression wave in such a material should have a rectangular profile.

The hardening law is usually described by empirical relations of the form.

$$\tau = \tau_0(1+A\gamma_p), \tag{3.14}$$

$$\tau = \tau_0 + B\gamma_p^n \tag{3.15}$$

where the hardening index is $n < 1$, γ_p is the plastic component of shear strain. Differentiation (3.14) gives

$$\frac{d\tau}{d\gamma_p} = \frac{\tau_0 n}{\left(1 + A\gamma_p\right)^{1-n}} \to \tau_0 n \text{ at } \gamma_p \to 0. \tag{3.16}$$

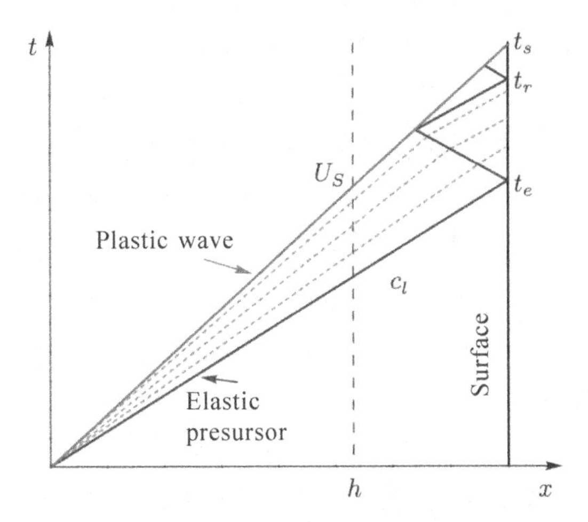

Fig. 3.3. Characteristics of an elastoplastic compression wave in a strain-hardening material.

That is, in this case there is a kink in the deformation diagram during the transition from elastic to plastic deformation. As a result, the elastic precursor takes the form of a jump with a region of constant parameters between its front and the front of a plastic shock wave. The effect of hardening according to (3.14) manifests itself in an increased plastic shock wave velocity.

By differentiating (3.15) with $n < 1$, we get

$$\frac{d\tau}{d\gamma_p} = \frac{nB}{\gamma_p^{1-n}} \to \infty \text{ at } \gamma_p \to 0. \tag{3.17}$$

This means the continuity of the derivative with the onset of plastic deformation is retained. In this case, the region of the anomalous curvature of the uniaxial compression diagram leads to the formation of a dispersion section behind the front of the elastic precursor, in which there is a continuous decrease in wave velocities as the stress increases.

The dispersion region is a simple wave described in the distance–time diagram (Fig. 3.3) with a fan of characteristics for which

$$d\sigma_x = -\rho_0 c_\sigma^2 d\varepsilon_x, \quad d\tau = \frac{3}{4}\left(1 - \frac{c_b^2}{c_\sigma^2}\right) d\sigma_x, \tag{3.18}$$

where c_σ is the phase velocity of propagation of a fixed stress level σ_x in Lagrange coordinates [8]. From here

$$\frac{d\gamma_p}{d\tau}=-\frac{d\varepsilon_x}{d\sigma_x}\frac{d\sigma_x}{d\tau}-\frac{1}{G}=\frac{4}{3}\frac{1}{\rho_0\left(c_\sigma^2-c_b^2\right)}-\frac{1}{G}, \tag{3.19}$$

$$\frac{d^2\gamma_p}{d\tau^2}=-\frac{8}{3}\frac{c_\sigma}{\left(c_\sigma^2-c_b^2\right)^2}\frac{dc_\sigma}{d\tau}=-\frac{32}{9}\frac{c_\sigma^3}{\rho_0\left(c_\sigma^2-c_b^2\right)}\frac{dc_\sigma}{d\sigma_x}. \tag{3.20}$$

Differentiation (3.15) gives

$$\frac{d\gamma_p}{d\tau}=\frac{\gamma_p^{1-n}}{Bn};\quad \frac{d^2\gamma_p}{d\tau^2}=\frac{1-n}{B^2n^2}\gamma_p^{1-2n}. \tag{3.21}$$

Equating the expressions for the derivatives (3.19) and (3.21) we get

$$\frac{4}{3\rho_0\left(c_\sigma^3-c_b^3\right)}-\frac{1}{G}=\frac{\gamma_p^{1-n}}{Bn};\quad -\frac{32}{9}\frac{c_\sigma^3}{\rho_0\left(c_\sigma^2-c_b^2\right)}\frac{dc_\sigma}{d\sigma_x}=\frac{1-n}{B^2n^2}\gamma_p^{1-2n}. \tag{3.22}$$

From this it follows that for $\gamma_p = 0$, $c_\sigma = c_1$ and for $n < 1$, the value of c_σ monotonously decreases with increasing plastic deformation. Considering that for a simple centred wave

$$c_\sigma=\frac{h}{h/c_l+t(\sigma_x)},\quad \frac{dc_\sigma}{d\sigma_x}=\left(\frac{\partial c_\sigma}{\partial\sigma_x}\right)_h=\frac{c_\sigma^2}{h\cdot(d\sigma_x/dt)_h}, \tag{3.23}$$

where h is the distance travelled by the wave, t is the time interval between the front of the precursor and the moment of reaching this stress level σ_x, we find that for $n < 1/2$ after an elastic shock, the stress gradient $\partial\sigma_x/\partial t \to \infty$, and with subsequent deformation the slope of the profile $\sigma_x(t)$ in the precursor decreases. In the case of $n > 1/2$, the value of $\partial\sigma_x/\partial_t$ beyond the elastic shock is zero and increases with plastic deformation. For $n = 1/2$, $d\sigma_x/dt = \text{const} \neq 0$. In this case, the hardening constant is determined by the relation

$$B^2=-\frac{9}{16}\frac{\left(c_\sigma^2-c_b^2\right)^3\rho_0}{c_\sigma^3\cdot dc_\sigma/d\sigma_x}. \tag{3.24}$$

3.3. The evolution of an elastic precursor in a relaxing material

Due to the limited speed of movement and multiplication of carriers of plastic deformation (dislocations and twins), the flow stress increases with increasing strain rate. Phenomenologically, the dependence of the flow stress on the strain rate is interpreted as a manifestation of 'viscosity' or stress relaxation in a solid. The dynamics of deformation of relaxing media is described by various models of elastic–viscous and elastic–viscous–plastic bodies. The simplest of these is the Maxwell model, which includes successively elastic G and viscous η elements (Fig. 3.4 *a*).

The total strain γ in the Maxwell model is the sum of the elastic γ_e and plastic (viscous) γ_p components:

$$\gamma = \gamma_e + \gamma_p. \qquad (3.25)$$

The elastic component of the deformation is related to the stress by the Hooke law: $\gamma_e = \tau/G$, while the stress in the viscous element is determined by the rate at which this deformation component changes: $d\gamma_p/dt = \tau/\eta$, where η is the viscosity coefficient. With the instantaneous application of a load, the deformation at the first moment is localized in an elastic element, then a viscous deformation develops, accompanied by stress relaxation, which is described by

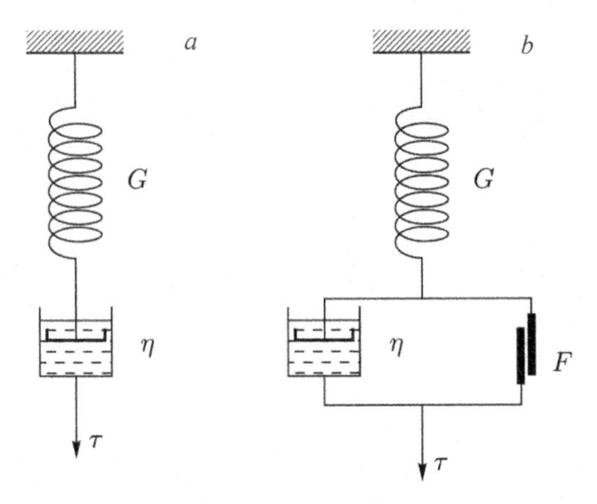

Fig. 3.4. Rheological models of viscoelastic deformation. G is an elastic element, η is an element with viscosity, F is a friction element.

the equation

$$\frac{1}{\tau}\frac{d\tau}{dt} = -\frac{G}{\eta}.$$

Integration of this equation leads to an exponential law of stress relaxation with time at a fixed total strain γ:

$$\tau = \tau_0 e^{-\frac{G}{\eta}t}. \tag{3.26}$$

The ratio of the coefficient of viscosity to the shear modulus, η/G, in the framework of this model is the relaxation time – a parameter often used as a characteristic of an elastic–viscous medium.

In another characteristic case of a fixed strain rate $\dot{\gamma}$, the stress monotonically increases from zero to a certain limiting value, determined by the coefficient of viscosity and the strain rate:

$$\tau = \eta\frac{d\gamma}{dt}\left[1-\exp\left(-\frac{G}{\eta}t\right)\right]. \tag{3.27}$$

The elastic–viscous nature of the deformation of a solid leads to the appearance of a number of specific features of the evolution of shock load pulses. The elastic precursor of a shock wave in such an environment decreases its amplitude as it propagates. For an elastic-viscous waves, the formation of relaxation zones, in which final states are asymptotically achieved, is immediately behind the regions with large parameter gradients.

Let us consider the decay of the precursor of a compression wave in an elastic–viscous medium in more detail, for which purpose we write down the partial derivatives of the particle velocity and stress at its front along the *h, t*-trajectory of its propagation [9]. Taking into account the laws of conservation of mass and momentum, we obtain:

$$\left.\frac{d\sigma}{dt}\right|_{\text{HEL}} = \dot{\sigma} - \rho_0 c_l \dot{u}, \quad \left.\frac{du}{dt}\right|_{\text{HEL}} = \dot{u} + \rho_0 c_l \dot{V}, \tag{3.28}$$

where c_l is the velocity of the elastic precursor, the dot above the symbol denotes the partial derivative with respect to time. Hence, in view of the Rankin–Hugoniot relations

$$2\frac{d\sigma}{dt}\bigg|_{\text{HEL}} = \dot{\sigma} + \rho_0^2 c_i^2 \dot{V}. \tag{3.29}$$

Let the shear stress relaxation be described by some function $F = G\dot{\gamma}_p$. Then

$$\dot{\tau} = -G\rho_0 \dot{V} - F, \ \ \dot{\sigma} = -\rho_0 E' \dot{V} - \frac{4}{3}F, \tag{3.30}$$

which, after appropriate substitution, leads to the well-known [6] equation for the decay of a precursor in a relaxation material with linear compressibility

$$\frac{d\sigma}{dt}\bigg|_{\text{HEL}} = -\frac{2}{3}F. \tag{3.31}$$

The last equation contains no flow parameters, except the stress at the front of the elastic precursor. In other words, in a relaxing linearly compressible material, the precursor should be attenuated regardless of whether the stress behind the shock on its front decreases or increases. In the case of a precursor having the form of a stress peak, the nonlinearity of the compressibility of the material also contributes to its attenuation.

The possibilities of applying the model of an elastic–viscous body are extended by introducing a friction element parallel to a viscous element (Shvedov–Bingham body, Fig. 3.4 *b*) and nonlinear viscosity.

In order to reveal the details of the process of the formation of an elastic precursor of a shock wave, consider for example the process of the formation of an elastic precursor after the collision of two plates of the same material. Due to the symmetry of such an impact, the velocity of the impact surface remains unchanged ($\dot{u} = 0$) until the beginning of unloading. In this case, at the initial stage

$$\dot{\sigma} = \frac{d\sigma}{dt}\bigg|_{\text{HEL}} = -\frac{2}{3}V \quad \text{and} \quad V = \frac{\sigma}{\rho_0 E'}. \tag{3.32}$$

In order to identify the trend of further changes in the wave profile, we consider the equation for the evolution of stress gradients and particle velocity directly behind the jump at the front of the elastic precursor:

$$\frac{d\dot{\sigma}}{dt}\bigg|_{\text{HEL}} = \frac{d\dot{\sigma}}{dt} - \rho_0 c_l \frac{\partial u}{\partial t}. \tag{3.33}$$

Since in the case of a symmetric impact, the velocity of the impact surface remains unchanged, $\partial u/\partial t = 0$ and $\partial \dot{u}/\partial t = 0$, then

$$\frac{\partial \dot{\sigma}}{\partial t} = -\frac{2}{3}\dot{F} \quad \text{and} \quad \left.\frac{d\dot{\sigma}}{dt}\right|_{\text{HEL}} = -\frac{2}{3}\dot{F}. \tag{3.34}$$

Thus, the stress gradient behind the precursor front should remain constant at the initial stage if the plastic strain rate does not change ($\dot{F} = 0$), form a peak if plastic deformation accelerates ($\dot{F} > 0$), and form a rise in the case of slowing down stress relaxation .

While an unambiguous evidence of stress relaxation behind the front of an elastic precursor is its decay as it propagates, the formation of a stress peak in the leading part of the precursor is evidence of the intense multiplication of carriers of plastic deformation – dislocations. The so-called 'sharp yield point' on the quasistatic deformation diagram cannot by itself lead to the formation of an elastic precursor wave in the form of a stress peak. The increase in the parameters behind the front of the elastic precursor can be a consequence of both strain hardening and slowing down stress relaxation.

3.4. Interpretation of the free surface velocity histories at the exit of the elastoplastic compression wave

Registration of free surface velocity histories $u_{fs}(t)$ (velocity of the back surface of the test material plate as a function of time) is currently the most common way to study the structure of intense elastoplastic compression waves in a solid. Although other measurement methods have been developed and are used, in particular, the registration of particle velocity profiles and compression stresses in the internal sections of the sample or on the border with a reference obstacle, measurements of the velocity profiles of the free surface have such advantages as simplicity of formulation, reliability and the highest temporal resolution. Very often, the interpretation of measurement results is limited to determining the value of the dynamic yield strength of a material, although the wave profiles contain more complete information about the deformation diagram $\sigma(\varepsilon)$ under compression and, therefore, about the strain hardening of the material, as well as its relaxation properties. However, it should be remembered that the velocity profile of the free surface is formed as a result of the interaction of the incident compression wave and the reflected rarefaction wave, and for this reason does not exactly reproduce the structure of the compression wave inside the test sample.

The deformation diagram is reconstructed from the measured profile of an elastoplastic compression wave within the approximation of a simple centred wave. For a simple wave described by a fan of rectilinear characteristics, the increments of the longitudinal stress $d\sigma_x$ and the strains $d\varepsilon_x = -dV/V_0$ are related by

$$d\sigma_x = \rho_0 a_\sigma^2 d\varepsilon_x, \quad (3.35)$$

where a_σ is the phase velocity of propagation of a section of a wave with a compressive stress σ_x in the Lagrange coordinates. The maximum shear stress τ under one-dimensional deformation in a shock wave is determined from the difference between the longitudinal stress σ_x and pressure p:

$$\tau = \frac{3}{4}(\sigma_x - p). \tag{3.36}$$

From consideration of the distance–time diagram for a simple centered wave, it follows that the phase velocity a_σ is defined as

$$a_\sigma = \frac{h}{h/c_l + t(\sigma_x)}, \tag{3.37}$$

where h is the distance between the impact surface (the pole of the fan of the characteristics of the centred wave) and the cross section in the sample for which the stress profile $\sigma_x(t)$ is analyzed, t is the time interval measured from the front of the elastic precursor. If instead of the stress profile $\sigma_x(t)$, the velocity profile of the free surface $u_{fs}(t)$ is analyzed, the empirical law of speed doubling is used:

$$u_{fs}(t) = 2u_p(t) \text{ and } d\sigma_x(t) = \rho a_\sigma \, du_p(t). \tag{3.38}$$

A more detailed analysis, taking into account the interaction between the incident and reflected waves near the sample surface, gives

$$a_\sigma = c_l \frac{2h - c_l t(\sigma)}{2h + c_l t(\sigma)}. \tag{3.39}$$

The plastic component of the strain γ_p is calculated by integrating the relationship

$$d\gamma_p = d\varepsilon_x - d\tau / G. \tag{3.40}$$

Since the compression time is known, the average strain rate in specific experiments is estimated quite accurately by dividing the deformation increment by the corresponding time interval.

The interaction of the elastoplastic compression wave with the free surface of the sample to be tested leads to the appearance of a series of reflections and distortion of the registered wave profile. Figure 3.5 shows the free surface velocity history obtained by computer simulation of the shock compression of a plate of an ideal elastoplastic material. It can be seen that although, in general, the graph is similar to the stress profiles in the internal sections of the plate, the free surface velocity history contains a number of additional features. The reasons for the appearance of the latter are illustrated by the stress–particle velocity and distance–time diagrams in Figs. 3.6 and 3.7.

The output of the elastic precursor to the free surface sets it in motion with a velocity $u_{fs,1}$, equal to twice the particle velocity at the Hugoniot elastic limit: $u_{fs,1} = 2u_{p,\mathrm{HEL}}$ and causes the appearance of a reflected rarefaction wave. After meeting the reflected wave with the plastic shock wave in the unloaded material, an elastic compression wave is again formed, which can be interpreted as a reflection of the elastic rarefaction wave from the plastic compression wave. This

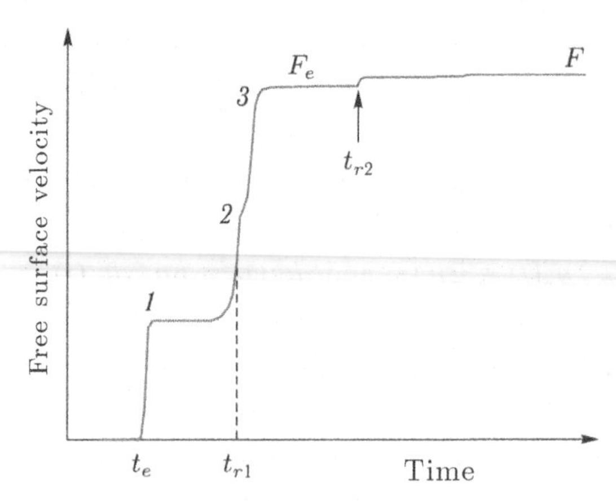

Fig. 3.5. The free surface velocity history of a plate of an ideal elastoplastic material, obtained as a result of computer simulation.

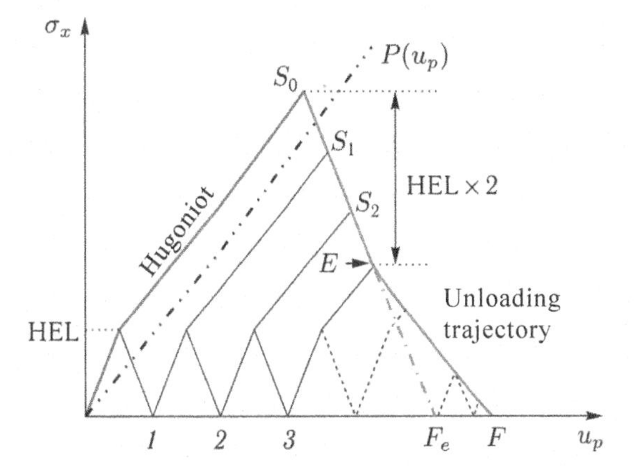

Fig. 3.6. The stress – particle velocity diagram of the interaction of an elastoplastic compression wave with the free surface of a plate. Digits 1, 2, 3 show the surface velocity values at the corresponding points of the wave profile in Fig. 3.5. Points S_0, S_1, S_2 show the initial state of shock compression and the states resulting from the interaction of the incident compression wave and the reflected rarefaction waves. Points F and F_e correspond to those indicated in Fig. 3.5.

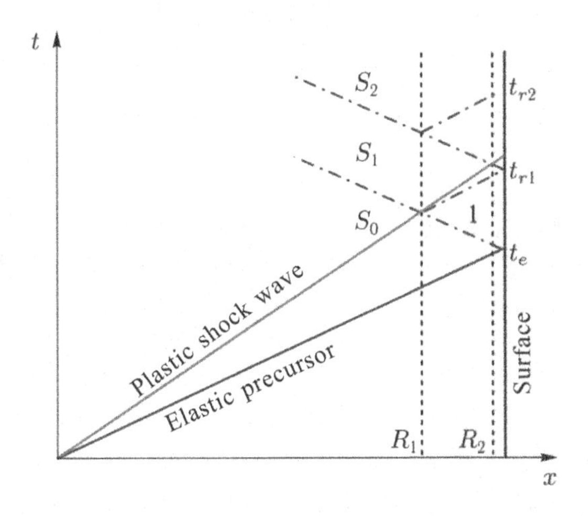

Fig. 3.7. The distance–time diagram of the reflection of an elastoplastic compression wave from the free surface of the plate. Points S_0, S_1, S_2 correspond to the states marked in Fig. 3.6.

reflection forms the second step on the velocity profile of the free surface – point 2 in Fig. 3.5.

Let the initial state of shock compression correspond to the point S_0 on the initial Hugoniot of the material (Fig. 3.6). After the interaction of the reflected and incident waves, the further propagation of the latter is described by the Hugoniot 1–S_1, shifted relative to the initial velocity along the axis by $2_{up,\,\mathrm{HEL}}$. It is essential that the unloading of the shock-compressed substance from the point S_0 to the point S_1 is of an elastic nature. Then the reverberation of the elastic wave between the free surface and the plastic front causes a further decrease in the stress from the state S_1 to the point S_2, and again the transition occurs in an elastic way, because the point S_1 on the offset Hugoniot corresponds to the state of the shock-compressed substance above the overall compression curve that meets the yield condition. Reverberations of an elastic wave cease with a drop in the stress in the second wave to the magnitude of the Hugoniot elastic limit.

During the reverberations, the state in the incident compression wave gradually passes from point S_0 to point F_e in Fig. 3.5 along the elastic unloading trajectory S_0–F_e. In other words, the thin near-surface layer undergoes only elastic deformation. At the same time, more distant layers located on the left side of the line R_1 in Fig. 3.7, were in a shock compressed state well above the elastic limit and should be unloaded in an elastoplastic manner along the trajectory S_0–E–F. As a result of this process, the speed of the free surface must be higher and correspond to the point F on the stress–particle velociy diagram. The mismatch of these apparent impedances of different layers of the plate causes additional reflections, which manifest themselves in the form of long steps in the upper part of the free surface velocity profile in Fig. 3.5. In the case of wave dispersion, this part of the free surface velocity profile of an elastoplastic body acquires a certain inclination. Thus, even in the absence of relaxation processes, when a plateau of stress occurs behind the plastic shock wave inside the plate, additional steps appear on the velocity profile of the free surface, which form the slope of the initial portion of the expected plateau of state parameters.

3.5. Plastic shock wave structure

While in classical gas dynamics the shock wave is usually represented as a discontinuity in the state parameters, and its width appears to be negligible, the rise time of the parameters during the shock-wave compression of a solid in many cases turns out to be quite measurable. The width of the plastic shock wave is determined by the relaxation time of shear stresses – a parameter inversely proportional

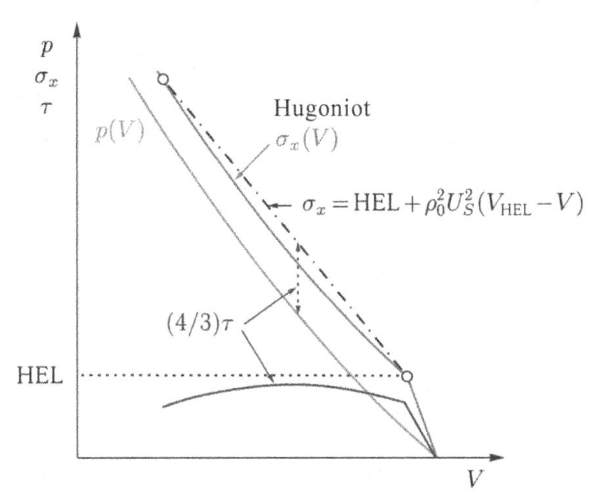

Fig. 3.8. The evolution of stresses in an elastic–plastic compression wave.

to the viscosity of the material. Swegle and Grady [10] found that the maximum strain rate $\dot{\varepsilon}_m$ in a plastic shock wave for various materials is related to the peak shock pressure by the universal relation:

$$\dot{\varepsilon}_m = A(\Delta\sigma)^4, \tag{3.41}$$

where A is the material constant, $\Delta\sigma$ is the difference between the maximum compressive stress in a plastic shock wave and the stress in the elastic precursor. In other variants, instead of $\Delta\sigma$ in the relation (3.41) we use the final compression stress behind the shock wave. If the compression wave is steady, then the change in deviator stresses in it is completely determined by the mutual position of the Hugoniot and the Rayleigh line, as shown in Fig. 3.8. In this case, the strain rate is determined by the kinetics of motion and multiplication of dislocations and is automatically set to be what it should be at given stresses and the prior history of deformation.

It can be shown that the maximum shear stress in a steady wave is approximately proportional to the square of the maximum shock stress. At the maximum point ($\dot{\tau} = 0$), the plastic strain rate $\dot{\gamma}_p$ is equal to the total compression rate $\dot{\varepsilon}_x$. In this case, the empirical Swegle and Grady ratio is converted to

$$\dot{\gamma}_p = A'\left(\tau - \frac{1}{2}Y\right)^2. \tag{3.42}$$

The value of the coefficient A' is approximately 10^8 GPa$^{-2} \cdot$ s^{-1} for aluminium, $6 \cdot 10^8$ GPa$^{-2} \cdot$ s^{-1} for bismuth, $3 \cdot 10^8$ GPa$^{-2} \cdot$ s^{-1} for copper, $3 \cdot 10^7$ GPa$^{-2} \cdot$ s^{-1} for iron, $5 \cdot 10^7$ GPa$^{-2} \cdot$ s^{-1} for beryllium. It follows from the last relation that the viscosity of solids is not a constant and decreases with increasing strain rate.

In the case of multicomponent composite materials, an additional factor leading to an increase in the shock wave width is the acoustic interaction between the components. The Swegle and Grady relation generally speaking, is not universal. Recent measurements of shock waves in glycerol have shown that in this case the dependence of the compression rate on pressure is much weaker. The reason is obviously related to the different nature of the viscosity of solids and liquids.

3.6. The formation of a two-wave structure during polymorphic transformation in the process of shock compression

The change of the crystal structure (polymorphic transformation) during the compression of crystalline materials is usually associated with an increase in their density. As a result, an anomalous compressibility region appears on the Hugoniot, which means the loss of stability of the shock wave and its splitting into two successive shock compression waves.

The effect of polymorphic transformations with volume change on the profiles of compression and rarefaction waves is illustrated in Fig. 3.9. The transition to a more dense phase begins at point 1 at pressure p_1 and ends at point 2 at pressure p_2. Region 1–2 is the area of mixed phases. Region 1–3 below the Rayleigh line passing through point 1 is inaccessible for single shock-wave transitions under compression from the initial state 0. Shock waves with pressure $p_1 < p < p_3$ split into two steady compression waves, with the first wave velocity $U_{S1} = V_0\sqrt{p_1/(V_0 - V)}$ is greater than the second-wave Lagrangian velocity $U_{S2} = V_0\sqrt{(p_0 - p_1)(V_0 - V)}$. Qualitatively, the situation is similar to the case of the loss of stability of a shock wave during an elastoplastic transition, but for rarefaction waves there is no such similarity.

When unloading the compressed high-pressure phase, the change in its state before the onset of the reverse transformation at point 2 corresponds to the unloading isentrope of this phase. The pressure

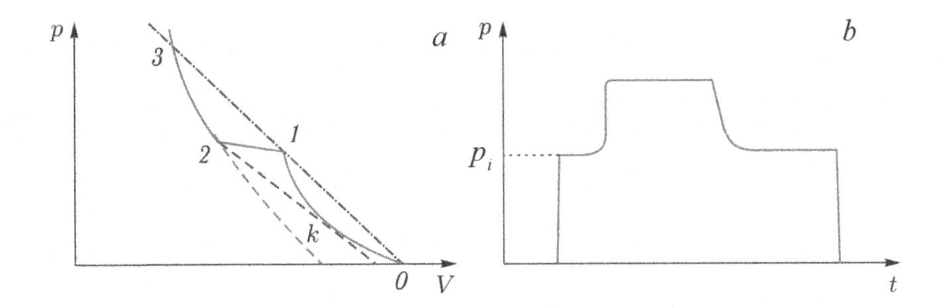

Fig. 3.9. The splitting of the shock wave and the formation of a shock rarefaction wave due to reversible polymorphic transformation with volume change. *a* is the Hugoniot of a material undergoing a polymorphic transition; *b* – pressure profile in a shock compression pulse.

range $p_k < p < p_2$ for rarefaction waves is anomalous in the sense that the speed of sound changes non-monotonically with decreasing pressure and is higher at point *k* than in section 2–1. As a result, during unloading, a rarefaction shock wave is formed, which propagates with a velocity U_R:

$$U_R = V_0\sqrt{(p_2 - p_k)/(V_k - V_2)}. \tag{3.43}$$

The formation of a two-wave compression configuration and a shock rarefaction wave is the most obvious and convincing evidence of a reversible polymorphic transformation in a compression pulse. In addition to the pressure of direct and reverse transformations, measurements of wave profiles give information about their rate. In this case, the kinetic data are obtained from the analysis of the decay of the first wave and measurements of the rise time in the second wave, depending on the shock compression pressure, just as it is done in the study of the kinetics of high-rate elastoplastic deformation.

3.7. Spall fracture of solids. Wave interactions during spalling

The dynamic strength of materials in the range of extremely small load durations is investigated by analyzing the so-called ‘spall’ phenomena at the reflection of compression pulses from the free surface of the body. The interference of the incident and reflected waves inside the body results in the formation of tensile stresses which can lead to its destruction and separation of the spall plate.

This phenomenon is called spallation or spalling. Resistance to the fracture of material in the conditions of spalling is called spall strength. The term 'spall strength' is not a strict physical definition and often causes objections because the value of the fracture stress is not constant, increases with an increase in the tension rate and is determined by the balance between the fracture rate and the load application rate rather than a certain 'failure threshold'. However, this term is adopted and used in the literature as a standard term for describing the dynamic tensile strength of materials.

Methods for measuring the resistance of materials to spall fracture are based on the analysis of wave interactions. Figure 3.10 shows the distance–time and particle velocity–pressure diagrams illustrating wave interactions upon the reflection of a triangular compression pulse from the free surface of the body. The exit to the free surface of the shock wave causes an abrupt increase in the surface velocity to a value u_0 equal to twice the particle velocity in the shock wave. The reflection of the shock wave from the surface causes the appearance of a centred rarefaction wave, which in the $t - x$ diagram is represented by a fan of the C_--characteristics. Following the shock wave, an incident unloading wave emerges on the surface, described by the C_+-characteristics in the $t - x$ diagram.

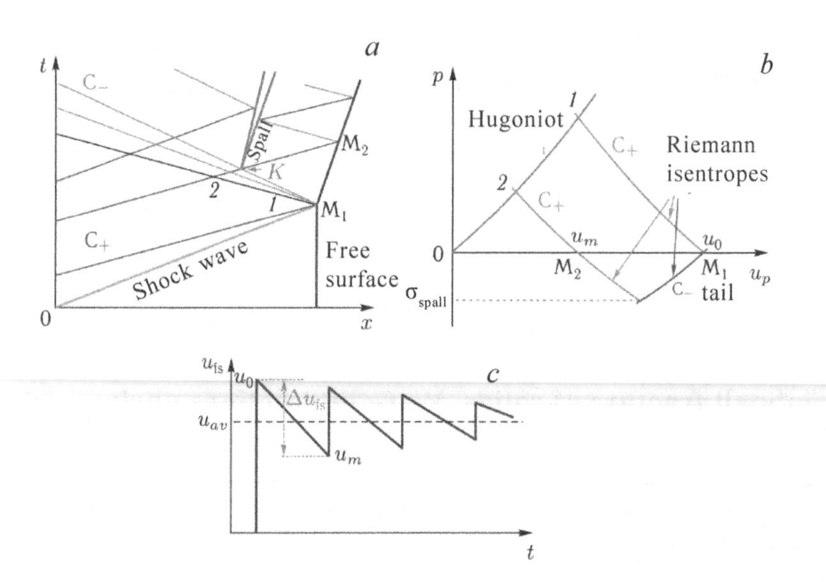

Fig. 3.10. Wave interactions during spallation: a – $t - x$ diagram of reflection of a compression pulse from the free surface; b - p – u diagram of wave interactions; c – velocity profile of the free surface during spall fracture.

A surge wave causes a decrease in surface velocity. The states of a substance in the zone of interaction between the incident and reflected waves are found in the *p–u* diagram as the intersection points of the Riemann trajectories of the state change along the C_+- and C_--characteristics passing through this point of substance at a given time. When two opposite opposite rarefaction waves interact, the pressure and particle velocity vary along curves parallel to the Hugoniot along the C_-- characteristics and, symmetrical to it, along the C_+-characteristics.

In each sample layer, the maximum values of tensile stresses are reached at the time of passage of the tail characteristic of the centred rarefaction wave. In particular, the state of the substance in the spallation plane immediately before the onset of fracture corresponds on the *p–u* diagram to the intersection point K of the Riemann trajectories M_1K (state change along the tail characteristic of the reflected centered wave) and $2K$ (the trajectory along the last C_+-characteristic of the unloading incident wave passing through the spalling plane). In the layer where the tensile stress exceeds a certain critical value for a given material, its fracture will occur – spalling takes place. If the fracture stress σ^* is fixed and does not depend on the time of action, and the failure occurs instantaneously, then the stress in the spall plane quickly relaxes from σ^* to zero (state on the free surface). As a result, compression waves will go to both sides of the spall surface. After the arrival of this compression wave to the free surface, its velocity again increases to u_0. Thus, a so-called 'spall pulse' is formed on the velocity profile of the free surface. Due to subsequent multiple wave reflections between the spall surface and the free surface of the sample, the motion of the latter occurs in the form of damped oscillations. In this case, the surface speed tends to an average value between the minimum and maximum speed. The amplitude of the waves reverberating in the spall plate decreases with time due to dissipative losses.

The slope of the spall pulse in real materials depends on the rate of fracture. If the fracture occurs without delay in a narrow region of the sample, then the rise time of the velocity in the front of the spall pulse is small. The spall fracture of materials with high viscosity is accompanied by the appearance of a weak spall pulse with a smooth increase and a rapid attenuation of surface velocity oscillations. Figure 3.10 shows schematically the typical velocity profile of the free surface of the sample during its spall fracture. Such free surface velocity history during the spallation were predicted by means of computer simulations and were observed experimentally in numerous experiments.

3.8. Determination of the magnitude of the breaking stress at spallation. Distortion of wave profiles at spalling in an elastic-plastic body

The magnitude of the fracture stress during the spallation (spall strength of the material) is determined on the basis of measurements of the free surface velocity history $u_{fs}(t)$. Relaxation of the tensile stress at failure leads to the appearance of a compression wave, the emergence of which on the surface of the body forms the so-called spall pulse on the profile $u_{fs}(t)$.

An analysis of the interaction of the incident and reflected waves by the method of characteristics gives the ratio between the stress in the spall plane σ^* and the surface velocity pullback Δu_{fs} from its maximum value in the compression pulse u_0 to the value u_m before the spall pulse front: $\Delta u_{fs} = u_0 - u_m$. In the linear approximation, this ratio has the form [11]

$$\sigma^* = \frac{1}{2}\rho_0 c_0 \Delta u_{fs}, \tag{3.44}$$

where ρ_0, c_0 are the density of the material and the speed of sound in it, respectively. Accounting for the nonlinearity of compressibility introduces a certain correction in (3.44), the value of which for spalling of millimeter thickness is usually small.

In the case of a clear manifestation of the elastoplastic properties of the test material, the question arises, which of the sound speed values should be used in (3.44): the speed of elastic longitudinal perturbations $c_l = \sqrt{[K + (4/3)G/\rho]}$ or 'bulk' sound speed $c_b = \sqrt{K/\rho}$ corresponding to the velocity of longitudinal waves in the field of plastic deformation.

Figure 3.11 shows a diagram of the longitudinal stress σ_x – particle velocity up for wave interactions when a compression pulse is reflected from the free surface of a body. Line H shows the Hugoniot, S_i, S_r – unloading trajectories in the incident and reflected rarefaction waves, respectively, R_{pl}, C_e — trajectories of the state change along the C_+ characteristics in the region of plastic tension before spalling and elastic compression after spalling. The slope of the initial part of the Hugoniot up to the elastic limit σ_{HEL} is $d\sigma_x/$

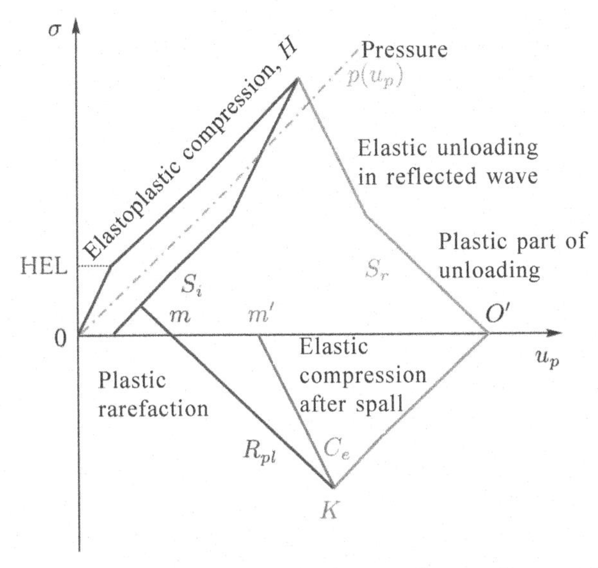

Fig. 3.11. The longitudinal stress σ_x–particle velocity u diagram for wave interactions when a compression pulse is reflected from the free surface of the body. H is the Hugoniot; S_i, S_r are the unloading trajectories in the incident and reflected rarefaction waves, respectively; R_{pl} and C_e are trajectories of state change along the C_+-characteristics in the plastic tension region before the spall and elastic compression after the spall.

$du = \rho c_l$. In the field of plastic deformation above the elastic limit $d\sigma_x/du = \rho c_b$. The rarefaction after shock compression also has an elastoplastic character. If the intensity of the shock compression pulse exceeds the value $2\sigma_{HEL}$, the tension in the interaction of the incident and reflected rarefaction waves is generated in the region of plastic strains.

In [12], attention was drawn to the fact that with the onset of fracture, plastic stretching in the detaching layer is replaced by its elastic compression. For this reason, the propagation velocity of the front of the spall pulse must be equal to the longitudinal sound velocity c_l, while the unloading part of the incident compression pulse in front of it propagates with a bulk velocity $c_b < c_l$. As a result, the free surface velocity history turns out to be distorted, and the velocity pullback in (3.44) $\Delta u_{fs} = u_0 - um'$ is underestimated compared to the value ($\Delta u_{fs} = u_0 - u_m$) expected in the neglect of the elastic–plastic properties of the solid. According to [12], the value of fracture stress is determined using the σ_x–u diagram (Fig. 3.11) at the intersection of the Riemann trajectory $O'K$ with a slope ρc_b corresponding to the tail C_--characteristic of the reflected rarefaction wave, and the trajectory $m'K$ corresponding to the C_+-characteristic of the front spall pulse. Point m' corresponds to the value of the velocity of the free surface ahead of the spall front. In this approximation

$$\sigma_c^* = \rho_0 c_l \Delta u_{fs} \frac{1}{1 + c_l / c_b}. \tag{3.45}$$

The relation (3.45), obtained by G.V. Stepanov [12], is for some reason called the Romanchenko relation in the English-language literature and is widely used to determine the spall strength. In this relationship, the thickness of the spall plate is not taken into account. Meanwhile, it seems obvious (and the available experimental data confirm this) that the distortion of the free surface velocity history of the free surface should depend on the thickness of the spall and the shape of the profile of the shock compression pulse. Accordingly, the use of the relation (3.45) in processing the data of experiments with varying loading conditions can lead to different values of strength, even if in fact its value is unchanged. In order to take this circumstance into account, an additional correction was introduced in the relation (3.45) in [13]:

$$\sigma_c^* = \rho_0 c_l \Delta u_{fs} \frac{1}{1 + c_l / c_s} + \Delta\sigma, \quad \Delta\sigma = \frac{1}{2} \frac{d\sigma}{dt}\bigg|_{C_-} h \cdot \left(\frac{1}{c_b} - \frac{1}{c_l} \right), \tag{3.46}$$

where $\left.\frac{d\sigma}{dt}\right|_{C_-}$ is the stress gradient along the tail C_--characteristic of the reflected rarefaction wave, equal to twice the gradient in the unloading part of the incident compression pulse, h is the thickness of the spall layer. The additional correction in [10] is not argued. In fact, it is assumed that with its introduction, the value of u_{fs} is determined, which would have occurred before the front of the spall pulse if the registered profile $u_{fs}(t)$ would not be distorted due to the difference in wave velocities. But then in the first relation (3.46) one should use the bulk sound speed, and not its combination with the longitudinal one. The corresponding relationship is:

$$\sigma^* = \frac{1}{2} \rho_0 c_b \left(\Delta u_{fs} + \delta \right). \tag{3.47}$$

The diversity in the methods for determining the spall strength with the same measurement method indicates that the analysis is incomplete and stimulates a more detailed consideration of wave interactions under the conditions of a spall in an elastoplastic body.

Figure 3.12 shows the results of computer simulation of the interaction of the 'catching-up' compression wave with the rarefaction wave in an elastoplastic body. At the left border of the plate, an impulse of shock compression of a triangular profile was initially created. After 1 μs, the linear decrease in the velocity of the boundary was replaced by its growth, which excited the second compression in the plate. In calculations, the steepness of the second compression wave was varied. The free surface velocity histories shown in Fig. 3.12, demonstrate the dependence of the propagation velocity of the front of the second wave on its steepness.

In order to obtain an expression for the velocity of the front of the second wave, we consider the distance x – time t diagram presented in Fig. 3.13. The diagram shows the C_+-characteristics of the original plastic rarefaction wave, followed by an elastic compression wave. Line F depicts the trajectory of the front of an elastic compression wave propagating at a speed c_F, which is in the range $c_l \geq c_F \geq c_b$. For the flow to the right of the trajectory F, the rate of change of stress along the trajectory, taking into account the conservation of the momentum equation, is expressed as

$$\left.\frac{d\sigma_x}{dt}\right|_F = \dot{\sigma}_+ - c_F \rho_0 \dot{u}_+, \tag{3.48}$$

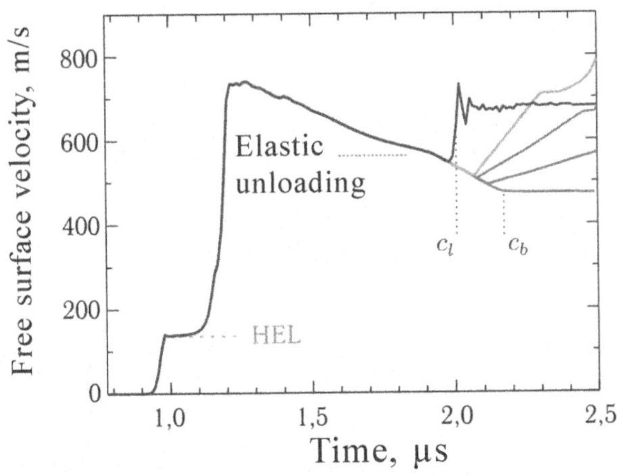

Fig. 3.12. The results of computer simulation of the interaction of the catch-up compression wave with the rarefaction wave in an elastoplastic body.

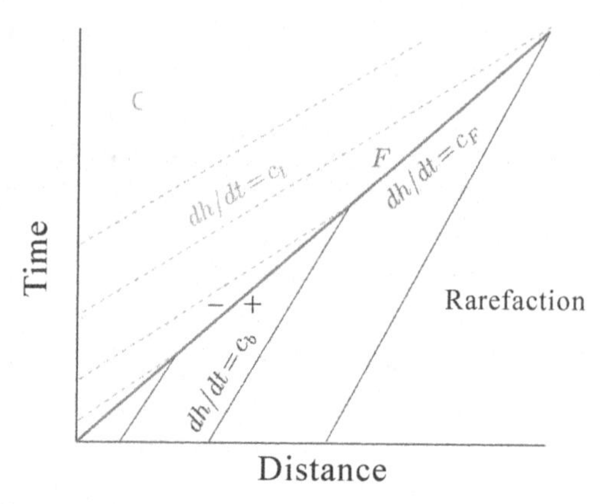

Fig. 3.13. The distance x – time t diagram of the interaction of the catch-up compression wave with the rarefaction wave in an elastoplastic body. The C_+-characteristics of the original plastic rarefaction wave, followed by the elastic compression wave, are shown. Line F depicts the trajectory of the front of a wave of elastic compression,

where the subscript '+' denotes parameters to the right of the trajectory F, the dot at the top indicates the time derivative: $\dot{f} = df/dt$. Similarly, to the left of the trajectory F

$$\left.\frac{d\sigma_x}{dt}\right|_F = \dot{\sigma}_- - c_F \rho_0 \dot{u}_-. \tag{3.49}$$

The particle velocity gradient to the right of the trajectory, taking into account the continuity equation, is expressed as

$$\left.\frac{du}{dt}\right|_F = \dot{u}_- + c_F \rho_0 \dot{V}_+. \tag{3.50}$$

Since it is assumed that the material before the front of the second wave is deformed plastically, then $\dot{V} = -\dot{p}/p^2 c_b^2$ Then we get

$$\left.\frac{du}{dt}\right|_F = \dot{u}_+ - \frac{c_F \dot{\sigma}_{z+}}{\rho_0 c_b^2}. \tag{3.51}$$

To the left of the trajectory F, plastic expansion is replaced by elastic compression, therefore the particle velocity gradient in this region is determined by the relation

$$\left.\frac{du}{dt}\right|_F = \dot{u}_- - \frac{c_F \dot{\sigma}_{x-}}{\rho_0 c_l^2}. \tag{3.52}$$

If there are no discontinuities in the flow, the stress and particle velocity gradients along the trajectory F on both its sides should coincide. Then we obtain two equations for the velocity of the front of the elastic compression wave, which catches up with the plastic rarefaction wave

$$c_F = \frac{\dot{\sigma}_{x+} - \dot{\sigma}_{x-}}{\rho_0 \left(\dot{u}_+ - \dot{u}_-\right)}, \quad c_F = \frac{\dot{u}_+ - \dot{u}_-}{\dot{\sigma}_{x+} / \rho_0 c_b^2 - \dot{\sigma}_{x-} / \rho_0 c_l^2}. \tag{3.53}$$

Excluding $\dot{u}_+ - \dot{u}_-$ from (3.53), we obtain the relation between the velocity of the second elastic wave and the stress gradients in the second wave and ahead of its front:

$$c_F = c_b c_l \sqrt{\frac{\dot{\sigma}_{x+} - \dot{\sigma}_{x-}}{\dot{\sigma}_{x+} c_l^2 - \dot{\sigma}_{x-} c_{b-}^2}}, \tag{3.54}$$

where $\dot{\sigma}_x +$ and $\dot{\sigma}_{x-}$ have different signs. In accordance with the solution obtained, the front of the catching-up compression wave propagates with the longitudinal sound speed only in two limiting cases: if the stress gradient in front of it is zero or if the catch wave is a shock discontinuity ($\dot{\sigma}_{x-} \rightarrow \infty$).

When a compression pulse of a triangular profile is reflected from the free surface of the body, the interference of the incident and reflected rarefaction waves occurs in such a way that a constant tensile stress is maintained in each section of the plate until the arrival of the spall pulse, that is, $\dot{\sigma}_{x+} = 0$. According to (3.54), the front of the spall pulse propagates in this case with a longitudinal speed of sound $c_F = c_l$ regardless of its steepness. In this case, the relations (3.47) and (3.45) for calculating the fracture stress are equally valid and give the same result, if the correction δ in (3.47) is calculated as

$$\delta = \left(\frac{h}{c_b} - \frac{h}{c_F}\right) \cdot \left|\dot{u}_1\right|, \tag{3.55}$$

where $c_F = c_l$. Usually, measurements of the spall strength are carried out when loading flat samples by the impact of a plate, which forms in the sample a shock compression pulse with a plateau of finite duration. In this case, the stresses in the sample sections in ahead of the spall pulse front are not constant; therefore, some of the characteristics of the elastic spall pulse front disappear in the process of interaction with the plastic rarefaction wave in front of it, as shown in Fig. 3.13. Consequently, the relation (3.45) is no longer valid, even if the spall pulse has a shock front. The calculation of the stress at the moment of spalling can be carried out using the relation (3.47) with the correction (3.55), where $c_F \neq c_l$ is determined by (3.54). For an idealized trapezoidal shock pulse, the value of c_F can be obtained by averaging its values, taking into account the fact that the free surface has $\dot{\sigma}_{x+} \approx 0$, while near the spallation plane $\dot{\sigma}_{x+} \approx \rho c_b \dot{u}_1/2$, $\dot{\sigma}_{x-} = \rho c_l \dot{u}_2/2$.

In all the approximations discussed, the quantities σ^* and δ are calculated under the assumption of almost instantaneous fracture concentrated in the spallation plane. In fact, the rate of fracture, due to the number of activated fracture nuclei and the rate of their growth, can not be arbitrarily large. Since the kinetics of destruction is not known a priori, it is impossible to be sure that the extrapolation of the sections of the $u_{fs}(t)$ profile used to estimate the value of δ is correct. For this reason, measurements of spall strength should be organized in such a way as to minimize the value of δ. The minimum distortion $u_{fs}(t)$ occurs at the triangular pulse profile of the shock load. The ratio between the measured rate decrement Δu_{fs} and the correction δt in (3.48) for a triangular load pulse is

$$\delta_t = \frac{1}{2}\Delta u_{fs}\left(\sqrt{\frac{3(1-\nu)}{1+\nu}} - 1\right). \tag{3.56}$$

where ν is the Poisson's ratio. With typical values of ν in the range from 0.3 to 0.35, the value of δ_t is 10–14% of the measured velocity pullback Δu_{fs}.

The loading of samples in experiments on the measurement of spall strength is usually carried out by the impact of a plate. In this case, the shock load impulse has at first approximately a rectangular shape until the distance covered by the compression wave reaches approximately five impactor thicknesses. After this, the rarefaction wave front catches up the compression wave, and the load impulse becomes close to triangular in shape. Therefore, in order for the correction value δ to be minimal, the ratio of the thickness of the sample and the impactor must be at least equal to 5.

In the nanosecond and picosecond ranges of duration, large tensile stresses are realized and the neglect of nonlinearity of compressibility introduces a large error in the determination of their values. In this case, when processing the measurement results, an extrapolation of the Hugoniot in the coordinates of compressive stress – particle velocity to negative pressures is used, which leads to the relation:

$$\sigma^* = \frac{1}{2}\rho_0 \left(c_0 - b\Delta u_{fs} / 2\right)\left(\Delta u_{fs} + \delta\right), \quad (3.57)$$

where c_0 and b are the coefficients of the linear expression for the Hugoniot $U_S = c_0 + bu_p$ (U_S is the velocity of the shock wave, u_p is the velocity of the particles of matter behind its front).

The magnitude of the resistance to spall fracture or the spall strength of the material characterizes the conditions for initiating fracture. As it develops, the material is 'weakened' by growing voids and the process ends at lower stresses. It is known that, at low load durations, the fracture, having begun, may not reach a complete breakaway, that is, until the body is divided into parts. To complete the process, additional energy costs are required, which are expended on the growth of nucleated voids and plastic deformation of the material around them. In this regard, the question arises of choosing the criteria for spall fracture, which would allow using the results of a limited number of material tests to predict its response under arbitrary loading conditions.

In the sixties and seventies, various empirical criteria were proposed that determined the possibility and completeness of spalling depending on the ratio of the amplitude and duration of the acting shock load pulse or energy storage in the spall plate. These criteria, however, do not take into account the real history of loading with stress relaxation in the process of fracture and are poorly compatible with computer simulation algorithms for shock-wave phenomena. Although criteria of this kind have been proposed so far, in the early seventies, instead of the finite criteria, it was proposed to use kinetic relations that determine the fracture rate as a function of the current value of tensile stress, the degree of fracture achieved, and other state parameters. With this approach, the spall strength in one way or another characterizes the stress at which the growth of the voids compensates for the increase in tensile stress during wave interactions.

The correctness of determining the spall strength from the surface velocity histories $u_{fs}(t)$ was repeatedly confirmed by experiments with shock load intensities close to the spall strength. In these experiments, the spall pulse was not recorded when the amplitudes of the load were less than the spall strength and appeared on the velocity profiles with an increase in the load above σ^*. With a further increase in load, the value of $\Delta u_{fs} = u_0 - u_m$ was practically preserved.

4
Main results of investigations of spallation phenomena in different materials

4.1. Typical values of sub-microsecond strength of solids and liquids

The main purpose of studying spallation phenomena is to obtain information for constructing constitutive relationships and models necessary for calculating the effects of an explosion, high-speed impact and other intense impulse effects in a wide range of parameters. To date, information has been obtained on the dynamic strength properties of a wide range of technical metals and alloys, ceramics and glasses, polymers and liquids. In this section, we present only some typical or, conversely, exotic research results. In recent years, the focus has been on assessing the governing factors and studying the details of the mechanism of the phenomenon and their connection with the structure of the material in order to find new areas of application of the shock wave technique for solving problems of materials science, physics of strength and plasticity. At the same time, methodological aspects of measurements become increasingly important. In particular, despite the fact that the basic idea of measuring spall strength is simple, insufficiently detailed analysis of the phenomenon often leads to a substantial quantitative discrepancy between the data obtained by different authors. It is necessary to clearly understand how the conditions of shock-wave tests can affect the development of fracture and the resulting values of spall strength.

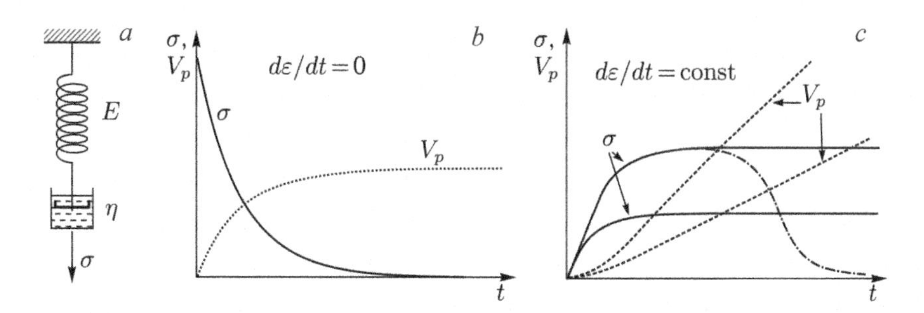

Fig. 4.1. To clarify, the stress relaxation within the Maxwell model (*a*) under instantaneous deformation (*b*) and deformation with a constant rate (*c*). V_p is the volume of voids in the material. The dash-dotted line on the graph (*c*) shows the effect of nonlinearity of the dependence of the resistance to growth of voids on their volume.

4.1.1. Loading conditions for measuring the spall strength of materials

In the practice of shock-wave testing of materials for the generation of uniaxial shock compression pulses, impact by a flyer plate, an explosive detonation in contact with a sample, and intense laser or corpuscular radiation pulses are used. In this case, the amplitude and duration of the action of the load, and the shape of the pressure profile (or mechanical stress) as a function of time, can vary within wide ranges. Since the fracture process is accompanied by stress relaxation, which in turn affects the further development of fracture, the real values of tensile stresses are associated with the loading history. Qualitatively, this relationship can be understood from consideration of the Maxwell model describing the behaviour of elastic–viscous media (Fig. 4.1 *a*). With instantaneous stretching, high stresses are realized in the model, which then gradually relaxes to zero as a result of the growth of voids in the material (Fig. 4.1 *b*). In the case of stretching with a constant rate (Fig. 4.1 *c*), stress relaxation due to the appearance and growth of voids begins at a lower stress and limits the growth of the latter.

Depending on the shape of the shock compression pulse generated in the test sample, the initial stage of spall fracture takes place in conditions close to one or the other of the two limiting cases shown in Figs. 4.1 *b* and 4.1 *c*. Figure 4.2 shows the loading history when a compression pulse is reflected from the surface with a triangular pressure profile. In this case, the tensile stresses first appear in the immediate vicinity of the free surface, their magnitude monotonously

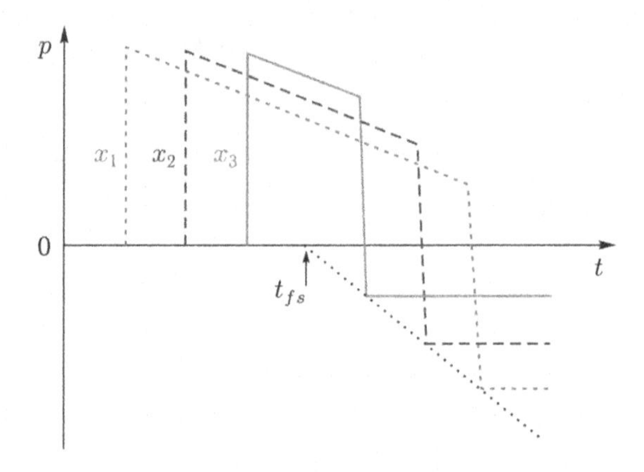

Fig. 4.2. The pressure profiles $p(t)$ in different sections x_i at the reflection of the compression pulse of a triangular shape from the free surface of the body. t_{fs} is the moment when the front of the compression pulse reaches the surface.

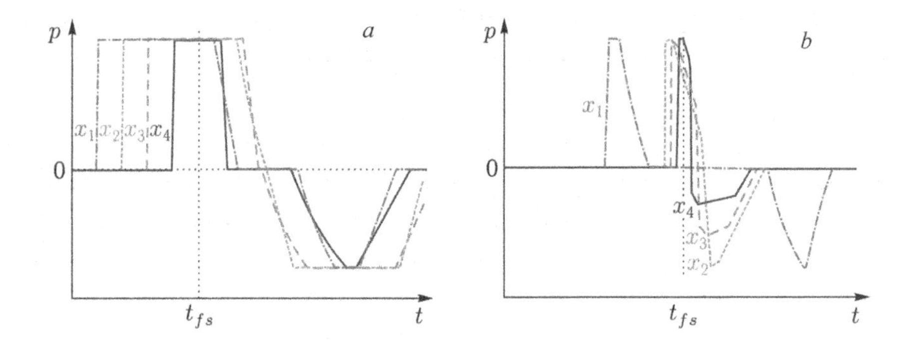

Fig. 4.3. *a* – pressure histories in several sections x_i of the target plate after the impact of the plate by half the thickness. The results of computer simulation. *b* – pressure profiles after impact by a thin flyer plate.

increases as the reflected wave propagates deep into the sample. In each selected section, the growth of the tensile stress occurs quickly, after which the stress is kept constant for some time, that is, the loading conditions close to those shown in Fig. 4.1 *b*.

In the practice of measuring the spall strength of materials, the generation of compression pulses by a plate is often used; the thickness of the plate is equal to half the thickness of the test sample, so that the spall fracture occurs close to the middle of the sample. Figure 4.3 *a* shows the loading histories in several sections of the target plate obtained for these conditions by numerical simulation with a real equation of state. In this case, the tension is first formed near the middle section of the sample, after which it propagates to the impact surface and to the back surface of the sample. In this case, due to the dependence of the speed of sound on pressure, the increase in tensile stress in all sections occurs for quite a considerable time, the situation is closer to that shown in Fig. 4.1 *c*. While in the case of a triangular impulse, it can be said that fracture begins to develop after achieving maximum tensile stress, then at a slower rate of application of the load, a noticeable fracture may occur even in the process of growth of the tensile stresses. To analyze the conditions for the initiation of fracture, it can be significant that the rate of increase of tensile stresses to the left of the middle cross section under these conditions is higher than to the right.

Figure 4.3 *b* shows the loading history in the case of a small (compared to the target) thickness of the impactor. With a sufficiently small ratio of the thicknesses of the impactor and the target near the back surface of the target, the shock compression pulse takes a form that is close to triangular, as a result of which load histories similar to those discussed for the triangular pulse are realized near the surface.

Figure 4.4 shows as an example the results of measurements of the free surface velocity histories of the samples of high-strength titanium alloy VT6. With a small amplitude of the load, the free

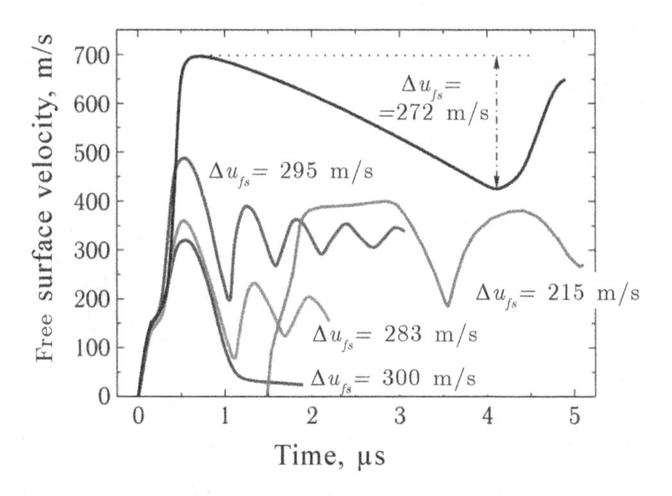

Fig. 4.4. The free surface velocity histories of flat samples of titanium alloy VT6, measured at different pressures and durations of shock compression pulses. The velocity pullbacks are indicated before the spallation of Δu_{fs}.

surface velocity histories practically repeat the shape of the shock pulse inside the sample. An exit to the surface of an elastoplastic compression wave followed by a rarefaction wave is recorded. With an increase in the shock load amplitude, tensile stresses also increase after the compression pulse is reflected from the surface, as a result of which the fracture process is initiated. As a result of the relaxation of stresses upon failure in a stretched material, a compression wave appears, which reaches the surface in the form of a spall pulse and again increases its velocity. Subsequent damped velocity oscillations are caused by multiple reflections of the waves between the surface of the sample and the spall surface. The velocity pullback before the spalling pulse, Δu_{fs}, is generally proportional to the magnitude of the tensile stress immediately before the onset of spall fracture, that is, the spall strength of the material. A further increase in the amplitude of the shock load pulse does not have a significant effect on the realized value of the spall strength. With an increase in the duration of the shock compression pulse, while maintaining the similarity of its shape, the resistance to spall fracture, as expected, decreases somewhat. Much bigger decrease of Δu_{fs} causes a transition from a triangular to a rectangular profile of the impulse of a shock-wave load. The latter is undoubtedly caused by the distortion of the wave profile due to elastoplastic effects, discussed in section 3.8.

As already mentioned, the rate of fracture, due to the number of activated foci of fracture and their growth rate, cannot be arbitrarily large. The measured values of the fracture stresses at spallation are a result of the competition between the growth of tensile stresses during wave interactions and their relaxation as a result of the appearance and growth of voids in the material. In other words, the resulting values of the fracture stress during spalling are controlled by the fracture kinetics. Consider this question in more detail.

Figure 4.5 schematically shows various options for the free surface velocity history, recorded in the study of spall phenomena. It is intuitively clear that processes accompanied by a fairly rapid relaxation of stresses can have a noticeable effect on the dynamics of wave interactions. Instant spalling forms a spall pulse with a shock jump at the front, while a slow, viscous fracture may not form a spall pulse, but only cause a slight deviation in the corresponding part of the free surface velocity history. Acoustic analysis of the process of the reflection of the compression pulse from the free surface of the body being destroyed [14] shows that the values of fracture stresses recorded at the minimum point ahead of the spall

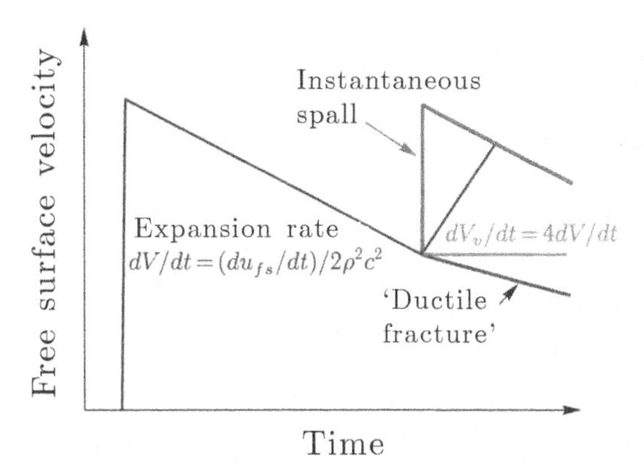

Fig. 4.5. The effect of the fracture rate on the spall pulse shape on the free surface velocity history.

pulse front correspond to the growth rate of the voids, approximately (with an accuracy of a constant of the order of several units) equal to the expansion rate of the material in the unloading part of the incident pulse.

Since the fracture kinetics is not known a priori, one cannot be sure of the extrapolation of the sections of the $u_{fs}(t)$ profile used to estimate the correction value δ. One more circumstance should be noted that significantly increases the error in determining the fracture stresses from experiments with a thick impactor (the ratio of the thickness of the impactor and the sample–target 1/2 used by Western researchers).

Figure 4.6 shows the pressure histories in samples of the D16 aluminium alloy measured with manganine gauges at two different impact speeds. Due to the elastoplastic nature of the discharge, the rate of decrease in pressure varies non-monotonously, which adds an

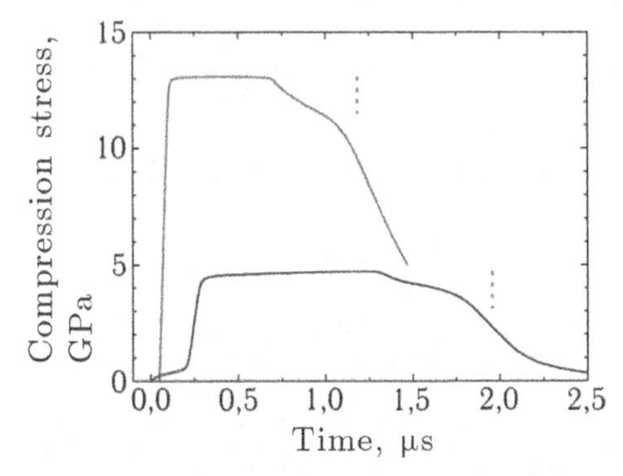

Fig. 4.6. Pressure histories in samples of D16 aluminium alloy at two different impact speeds.

extra uncertainty to the estimate of the correction value δ. In addition, since the yield strength of a material increases with pressure, the magnitude of the initial ('quasi-elastic') section of the rarefaction wave is greater, the greater the speed of impact. If the magnitude of the spall strength corresponds to the length of the vertical dashed segments in Fig. 4.6, then the magnitude of the velocity gradient $\dot{u}_1$ ahead of the spall pulse front and, accordingly, the correction δ turn out to be smaller in the case of a high impactor speed than in the case of a low impact speed. Probably, this circumstance is one of the reasons for the increase in the measured values of spall strength with increasing pressure of the preceding shock compression observed in experiments with thick impactors and not observed in experiments with thin impactors. We note that a satisfactory explanation of the dependence of the spall strength on the pressure of the preceding shock compression has not as yet been found.

Measurements of spall strength should be organized in such a way as to reduce the value of δ to a minimum. Analysis of the sources of error shows that this is achieved in the case when at the moment of reaching the surface the impact shock pulse has a profile of a triangular or close to it shape.

4.1.2. Spall strength of materials and substances of various classes

Figure 4.7 shows the results of experiments to determine the spall strength of aluminium in various structural states – technical aluminium AD1, high-purity polycrystalline aluminium, and aluminium single crystal. Homogeneous single crystals do not contain such relatively large potential foci of fracture, such as grain boundaries, inclusions, micropores, etc., and due to this they demonstrate the highest spall strength, 2.5 times higher than the strength of technical aluminium. High-purity polycrystalline aluminium has lower resistance than the single crystal, but significantly higher than the technical one, which contains up to 0.7% impurities.

Figure 4.8 compares the results of experiments with sapphire samples (single crystal alumina) [18] and ceramics sintered from alumina powder [19]. The maximum stress at sapphire shock compression was below its Hugoniot elastic limit. In these experiments, sapphire showed a spall strength of 8.9 GPa at a maximum compressive stress in the preceding shock wave of

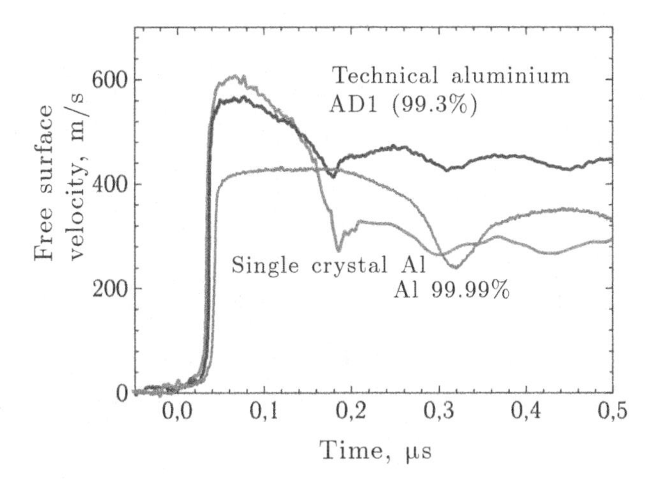

Fig. 4.7. The results of experiments to determine the total strength of technical aluminium AD1 [15], high-purity polycrystalline aluminium [16], and a single-crystal sample of aluminium [17].

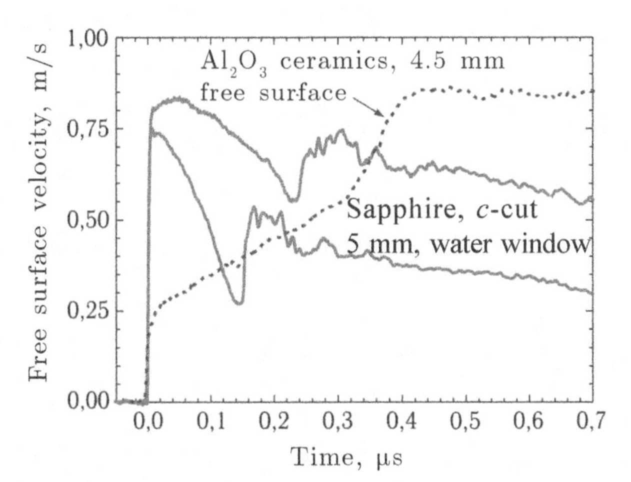

Fig. 4.8. Contact surface velocity histories measured in two experiments with sapphire and the the free surface velocity history of ceramic alumina samples.

18.2 GPa and 4.2 GPa after shock compression to 20.6 GPa and a longer duration of the load pulse. With a duration of ~40 ns and a compressive stress in a shock wave of 23 GPa, the measured spall strength of sapphire was 20 GPa. When the Hugoniot elastic limit is exceeded in the previous compression wave, inelastic strains lead to an almost complete loss of resistance to spalling of sapphire. Ceramic alumina, in which the grain boundaries are potential foci of fracture, has a spalling strength of about 0.5 GPa after shock

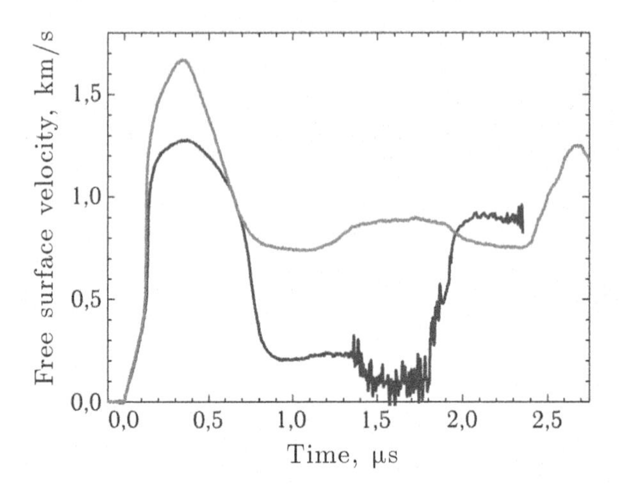

Fig. 4.9. The velocity profiles of the free surface of the LK7 glass plates with a thickness of 6 mm, measured on impact with an aluminium plate with a thickness of 2 mm and a speed of 1.2 km/s and 1.8 km/s. In the latter case, a thick paraffin layer followed the impactor.

compression, both below the elastic limit and substantially higher than it. This spall strength value is typical for ceramics and rocks. Figure 4.9 shows the results of experiments to measure the spall strength of borosilicate optical glass LK7 after shock compression above and below the dynamic elastic limit [20]. In either case, the spall was not observed, which means very high glass strength: more than 6.5 GPa below the elastic limit (which is in the range from 7.5 to 8 GPa) and remaining large with an excess of elastic limit. Obviously, unlike sapphire, inelastic deformation of glass occurs in a homogeneous way and is not accompanied by the appearance of stress concentrators.

Figure 4.10 shows the velocity profiles of the free surface of PMMA (polymethyl methacrylate) and rubber samples. PMMA is destroyed by the growth of lenticular cracks, which originate on inclusions of solid particles and gas pores existing in the source material. The shape of the wave profile at a low shock load intensity in this case is typical for solids. Typical values of the spall strength of solid polymers are in the range from 0.15 to 0.3 GPa with durations of shock-wave load of about 10^{-7}–10^{-6} s. Elastomers, in particular rubber, have the ability to undergo very large reversible strains without rupture. The behaviour of these materials during the spallation is somewhat different from the behaviour of other solids. The free surface velocity histories of the rubber sample in Fig. 4.10

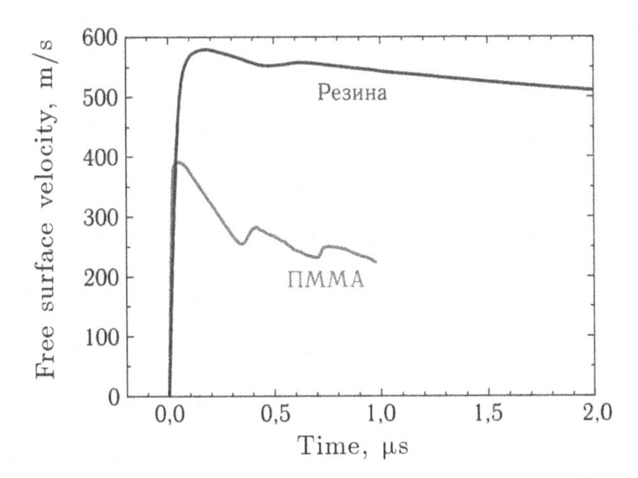

Fig. 4.10. The behaviour of plexiglass (PMMA) [21] and rubber [22] during spalling.

only in a small initial section corresponds to the incident shock pulse of compression; then a weak spall pulse is recorded. After the start of spalling, the detachable surface layer is braked for a long time, which, obviously, retains its connection with the main part of the sample. Inspection of the remaining rubber specimen did not reveal any noticeable damage. The measured spall strength of rubber was 27 ± 3 MPa, which is even less than the true value of rubber tensile strength S_k at low-speed stretching under the uniaxially stressed condition (88 MPa).

It is known that the rupture of elastomers is preceded by the formation of pores. The origin of these pores occurs at stresses much smaller than the breaking point. In the case of a stress state close to three-dimensional stretching, as is the case when exposed to a uniaxial compression pulse, the pores are generated after a small tensile strain. However, after this, the stress state between the pores is removed from the three-dimensional stretching, as a result of which large reversible strains of the porous rubber become possible. The measured value of the spall strength characterizes the conditions for the nucleation of micro-voids in rubber, but does not characterize the condition of its complete rupture. The formation of micro-breaks is not in itself a fracture. Thereafter, the elastomers can undergo significant stretching, which is completely or partially reversible. The protracted deceleration of the free surface after the appearance of the spall pulse is determined by the resistance to tension of the porous rubber.

Liquids, like solids, have resistance to three-dimensional stretching, which can be interpreted as bulk strength. Figure 4.11 shows the results of experiments [23] for measuring the spall strength of water and glycerol. The measured value of the spall strength of water was approximately 40 MPa, for glycerol it varies from 60 MPa to 140 MPa, depending on the strain rate. We note the atypical feature on the free surface velocity history during spallatioin in water (Fig. 4.11), which manifests itself in the form of a large amplitude and a short rise time of the spall pulse. The surface velocity in the spall pulse becomes even greater than the velocity immediately after the shock wave reaches the surface. It can be assumed that this excess is a consequence of the inertia effects of the growth of cavitation bubbles. The free surface velocity history during the spallation in more viscous glycerol does not have such an anomaly. Both wave profiles in Fig. 4.11 demonstrate a distinct delay of spall fracture, which is manifested in the fact that the period of velocity oscillations after spalling is substantially less than the duration of the first velocity pulse before spalling. As a rule, a similar delay for solids is not clearly recorded. Another possible, but less probable reason for the decrease in the period of oscillations of u_{fs} may be the development of cavitation processes in the breakaway layer of liquid after the spalling pulse is reflected from the surface.

Recently, it was possible to measure the spall strength of molten metals both in the submicrosecond range of shock load durations [24, 25] and in the picosecond range [26]. It turned out that at relatively

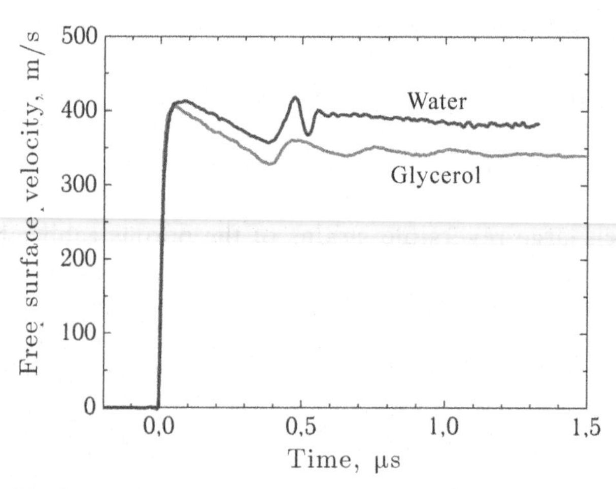

Fig. 4.11. The free surface velocity histories, measured in experiments on spallation in water and glycerol [23].

large times, the spall strength of liquid tin, lead, and zinc is an order of magnitude less than the strength of these metals in the solid state.

4.1.3. Approaching the 'ideal' strength of a condensed substance

Since the fracture at high overstresses in a material is a kinetic process of nucleation, growth and coalescence of pores or cracks and cannot occur instantaneously, an increase in the rate of load application allows one to create ever higher tensile stresses in the material. At present, it has become possible to carry out measurements at a level of stresses comparable to the ultimate possible or 'ideal' strength of condensed matter. It is known that the elastic moduli of solids (and liquids) increase with compression and, accordingly, decrease with tension. At a certain value of tensile stresses, the bulk modulus should vanish: $dp/dV = 0$, which corresponds to the absolute loss of stability of the condensed state. These stress values are understood by the term 'ideal strength'. There is a fairly strong correlation between the melting point and the magnitude of the ideal tensile strength [27].

It is interesting to compare the measured values of the spall strength of homogeneous condensed media with the maximum possible values of tensile strength. Figure 4.12 shows the dependences of the normalized σ_{sp}/σ_{id} values of the spall strength σ_{sp} of metallic single crystals, amorphous polymers and liquids on the strain rate. The values of the ideal strength σ_{id} were estimated as the pressure at the minimum of the Hugoniot of a substance extrapolated to the stretching domain:

$$\sigma_{id} = \frac{\rho_0 c_0^2}{4b}, \tag{4.1}$$

where c_0, b are the coefficients of the linear expression for the Hugoniot in the form $U_S = c_0 + bu_p$ (U_S is the velocity of the shock front, u_p is the particle velocity of matter behind the front). As shown by the comparison with the first-principle calculations shows, the error of such an estimate of σ_{id} is approximately 20% with a tendency to overestimate σ_{id}. For illustration in Table 4.1 the results of the estimations with (4.1) are compared with the results of determining σ_{id} from the first-principle calculations, extrapolation of the equation of state of matter (EOS) to negative pressure, and

Table 4.1. Comparison of various ways to assess the ideal strength

Substance	Estimate σ_{id} according to (4.1), GPa	σ_{id} *ab initio*, GPa	σ_{id} at 300 K from EOS or MD, GPa
Aluminium	13.4	11.7 [29]	12.2 – EOS [27]
Copper	23.3	19.8 [30]	21.2 – EOS [27]
Molybdenum	55	41.6 [41] 42.9 [30]	
Iron	31.6	27.9 [32] 27.7 [30] 13.4 [33]	13.5 – EOS [27]
PMMA	1.39		
Epoxy resin (polymerized)	1.34		
Water	0.28		0.21 – EOS [27] 0.22 – MD [35]

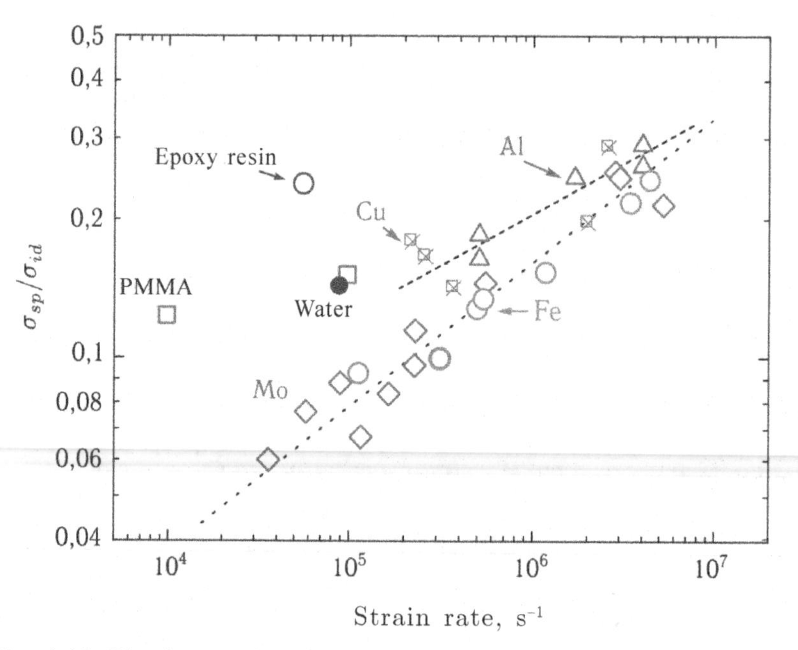

Fig. 4.12. The degree of realization of the ideal strength σ_{id} of homogeneous materials (single crystals, amorphous polymers and liquids) at spallation, depending on the strain rate.

by molecular dynamics (MD) simulations. Note that first-principle calculations are carried out for the temperature of absolute zero and as a result, give upper estimates of ideal strength, σ_{id} decreases with increasing temperature. Although the measured values of the spall strength of this range of materials differ by more than two orders of magnitude, in the normalized coordinates in Fig. 4.12 the variants of data are not so great.

In Fig. 4.12, the strain rate is understood as the rate of expansion of a substance in a rarefaction wave ahead of a spall pulse, which is defined as

$$\frac{\dot{V}}{V_0} = -\frac{\dot{u}_{fsr}}{2c_b}, \tag{4.2}$$

where $\dot{u}_{fsr}$ is the rate of fall of the velocity of the free surface ahead of the spall pulse. Although in reality the rate of tension in the interaction of counterpropagating rarefaction waves varies widely, the representation of experimental data in the form of Fig. 4.12 has the advantage of being directly used to evaluate fracture kinetics. It was shown [14, 28] that the initial rate of growth of the volume of voids during spallation with an accuracy of a constant factor of ~(2–4) is equal to the rate of expansion of the substance in the rarefaction wave, calculated in accordance with relation (4.2).

The data presented in Fig. 4.12 show that at nanosecond load durations up to 30% of the ideal strength of a condensed substance is realized. Ductile single crystals of copper and aluminium with the FCC structure demonstrate a slightly higher degree of realization of ideal strength than iron and molybdenum, having the BCC crystal lattice structure. This is probably due to the possibility of a higher stress concentration in the vicinity of microdefects in the BCC metals with a higher yield strength. The degree of realization of the ideal strength for spallation of amorphous polymers and liquids is at least not lower than that of metals. The difference in the degree of realization of the ideal strength of various substances decreases as the duration of the load decreases.

In recent years, it has become possible to carry out shock wave measurements in the picosecond range of load durations. Figure 4.13 summarizes the results of measurements and atomistic simulations of high-rate fracture and spalling and first-principle calculations of the 'ideal' strength of aluminium. Extrapolation of the experimental data to higher strain rates shows their agreement with molecular

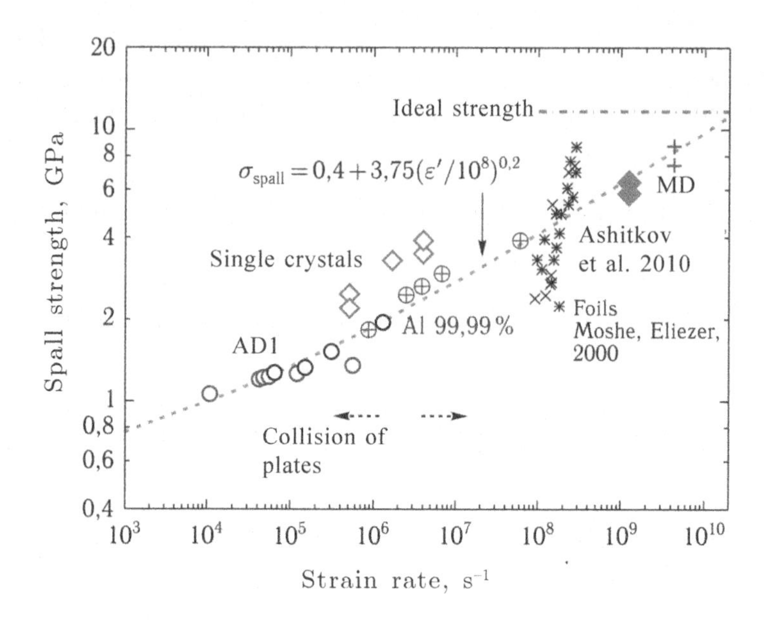

Fig. 4.13. The results of measurements of the spall strength of aluminium of different purities in comparison with the data for monocrystalline aluminium, the results of molecular dynamics simulation of spalling, and the value of the ideal strength of aluminium from the first-principle calculations. Measurement ranges with plate impact and lasers are shown.

dynamics calculations and predicts the achievement of 'ideal' strength at a stretching rate of approximately $2 \cdot 10^{10}$ s^{-1}.

Measurements have shown that the spall strength of liquid tin, lead, and zinc is a much smaller fraction of ideal strength than is the case for water and other liquids at room temperature. In the picosecond range, the spall strength of molten tin was 1.9 ± 0.3 GPa, that is, less than 30% of the ideal strength. For metals in the solid state, the spall strength measured in this range of duration usually exceeds 70% of the ideal strength.

4.1.4. The influence of structural factors on the sub-microsecond strength of metals

Figure 4.14 shows the results of measurements of the free surface velocity histories of samples of the D16T alloy under shock-wave loading in the rolling direction and in the transverse direction. With the chosen ratio of the thicknesses of the impactor and the sample, the loading conditions near the free back surface of the sample

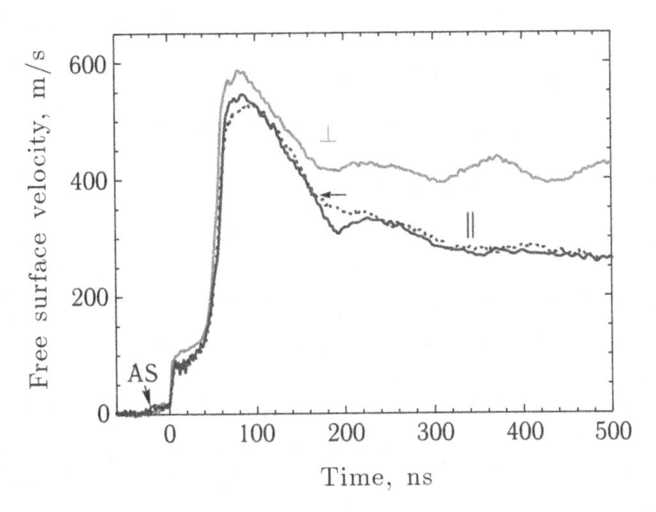

Fig. 4.14. The free surface velocity histories of D16T alloy samples with a thickness of 2 mm in the initial state under loading by a plate 0.4 mm thick. Symbols || and ⊥ denote the loading conditions in the rolling direction and in the transverse direction, respectively. For an impact in the direction of rolling, the results of two experiments are given. AS is the result of the impact of an air shock wave in front of the flyer plate. An arrow on the profile without a clear spall pulse indicates the expected onset of fracture.

correspond to the onset of decay of the shock wave under the action of the rarefaction wave catching it.

From the wave profiles presented on Fig. 4.14 it can be seen that the spall fracture of the alloy under loading in the direction of rolling has a prolonged character: for a long time after the onset of fracture, the splitting surface layer remains connected with the rest of the sample and, as a result, continues to decelerate. It should be noted that a similar type of fracture was previously observed, in particular, in experiments with 09G2S steel [36], loaded, as in our case, in the direction of rolling. The completion of the fracture is associated with the work of growth and coalescence of the nucleated voids, which consumes the kinetic energy of the coming off plate. The difference in the kinetic energies of the spall layer immediately before the fracture and after its completion is determined by the work of fracture.

Comparison of two wave profiles obtained under the same loading conditions shows that the recorded resistance to fracture during spalling under shock compression in the longitudinal direction varies from shot to shot. Variations of the spall strength values are obviously explained by the heterogeneity of the material structure. Such effects can be observed if the spatial resolution of the measurements (in this case, 0.1 mm) is less than the characteristic size of the transverse inhomogeneities of the velocity field. In addition, the effects of fracture inhomogeneity are smoothed as the signal propagates from the spallation plane to the free surface of the sample. The thickness of the spallation in these experiments was 0.35 mm, therefore, we can speak about the characteristic size of the heterogeneity of the structure of the material, greater or approximately equal to 0.1–0.3 mm. Since the time point of 250 ns after the wave front has reached the free surface, the results of the two experiments under discussion practically coincide.

Direct comparison of wave profiles in Fig. 4.14, obtained under loading conditions in the rolling direction and in the transverse direction, demonstrates a clearly lower resistance to spall fracture in the latter case. In this respect, the behaviour of duralumin D16T is similar to that previously observed for textured steels and alloys. At the same time, metallographic analysis of the stored steel samples [36] showed the absence or low degree of coalescence of cracks or pores formed in the direction perpendicular to the direction of the shock load. With the transverse orientation of the shock load, the fracture is completed faster, the braking of the spall plate is practically not observed, and the magnitude of the work of fracture is substantially less than when loaded in the direction of rolling.

Figures 4.15 and 4.16 show the results of experiments with Armco's iron in the as-rolled and ultrafine-grained states. In the as-rolled state, the average grain size was 28 μm. The ultrafine state of Armco iron was obtained as a result of severe plastic deformation by the method of forging with a change of the deformation axis with a total strain of more than 1500%.

After forging, the dimensions of grains decreased to such an extent that it was impossible to determine their average size by metallographic sections.

The data in Figs. 4.15 and 4.16 show that in this case a decrease in the grain size led to an increase in the magnitude of the spall strength of iron. However, this is not a general pattern. There are examples of the reverse effect of grain size on the spall strength of metals, that is explained by the increase in the total area of grain boundaries, which are potential nucleation sites for damage. The velocity profiles of the free surface also demonstrate an increase in the magnitude of the real resistance to fracture with a reduction in the duration of the shock load pulse.

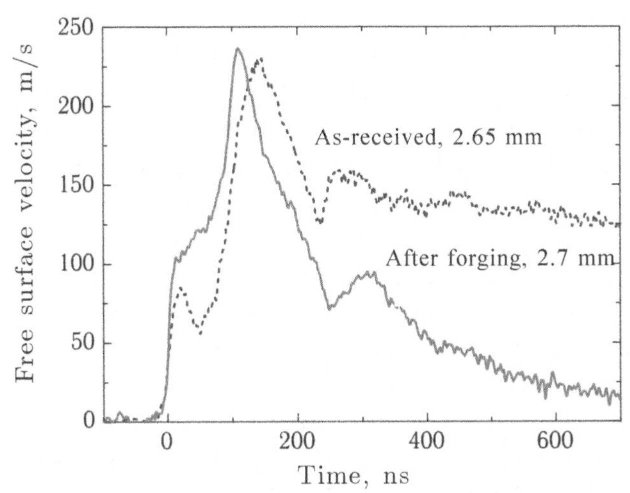

Fig. 4.15. The free surface velocity histories of Armco iron specimens with a thickness of 2.7 mm, loaded by an impact of an aluminium plate with a thickness of 0.4 mm and a speed of 600 ±30 m/s.

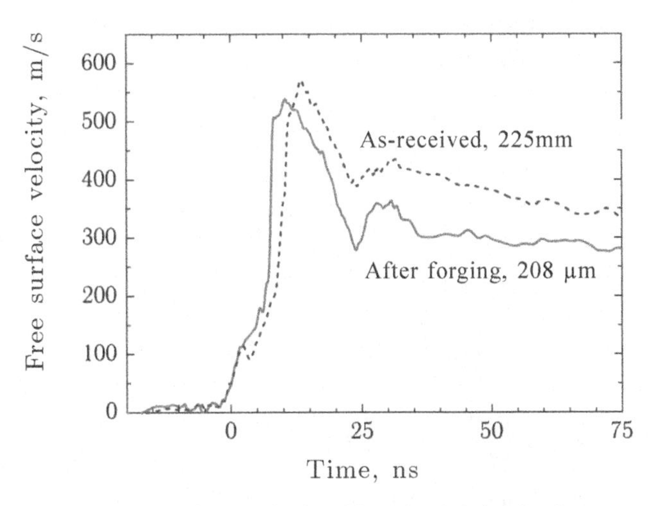

Fig. 4.16. The free surface velocity histories of Armco iron specimens with a thickness of 0.2 mm, loaded by an impact of an aluminium plate with a thickness of 0.05 mm and a speed of 1200 ± 50 m/s.

Figure 4.17 summarizes the results of measurements of the resistance to spall fracture of Armco iron in the as-received and ultrafine-grained states, as well as individual large (5–10 mm) grains, actually single crystals, high-purity iron, depending on the strain rate. The resistance to fracture of submicrocrystalline iron differs just little from the spall strength of single crystals.

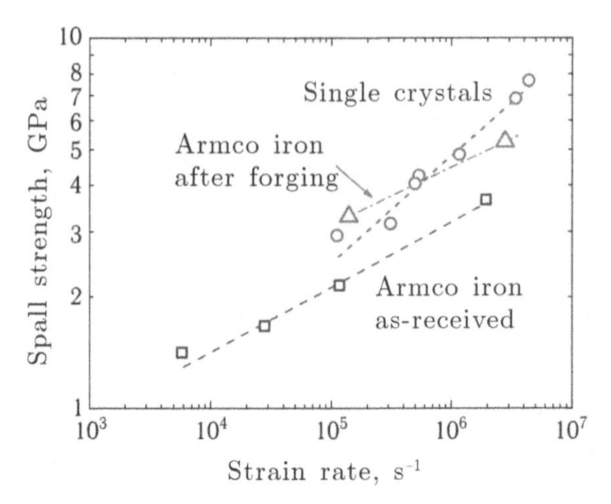

Fig. 4.17. The dependence of the resistance to spall fracture of iron in various structural states on the rarefaction rate.

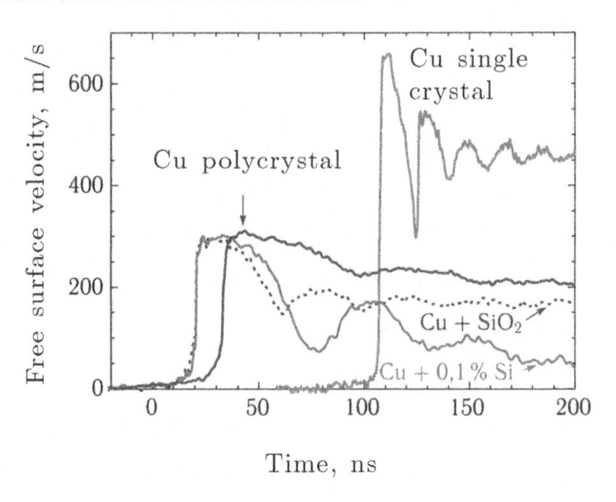

Fig. 4.18. The free surface velocity histories of single-crystal samples of copper 0.2 mm thick, copper with 0.1% silicon 0.5 mm thick and copper with inclusions of silicon oxide 0.5 mm thick, as well as a sample of high-purity polycrystalline copper 1 mm thick. The thickness of the aluminium impactors ranged from 0.05 mm to 0.2 mm in proportion to the thickness of the test specimen; impact speed was 1.2 ± 0.05 km/s and 0.66 ± 0.03 km, respectively.

Figure 4.18 presents the results of experiments with copper in the nanosecond range of shock wave loads. The samples were high-purity copper single crystal and polycrystalline states, a single crystal of copper with 0.1% silicon and a single crystal of Cu with 0.1% Si, heat-treated at 1030°C in cuprous oxide for 24 hours. It is known that silicon with copper forms a substitution solid solution, whose exposure in a Cu_2O powder at high temperature leads to the formation of scattered SiO_2 particles with a size of ~350 nm in a single crystal matrix.

The measurement results shown in Fig. 4.18 demonstrate a clear difference in the values of the spall strength and the nature of the spall fracture depending on the structural state of the tested samples. A pure copper single crystal has the highest strength and is characterized by a rapid completion of the fracture process. The surface speed of the splitting plate oscillates due to wave reverberation, but its mean value is quickly set to be constant, which means that the plate braking terminated 5–10 ns after the start of the spall. Polycrystalline copper has the least resistance to fracture, but the process of fracture develops slowly. In this case, the connection between the splitting layer and the rest of the sample is maintained for a relatively long time, which follows from the observed prolonged deceleration of the spall plate. The single crystal of the Cu + 0.1% Si solid solution has a slightly smaller strength than the pure copper single crystal, and exhibits a viscous, prolonged fracture pattern. With the formation of fine particles of SiO_2, a further decrease in the resistance to spall fracture and its acceleration are associated.

Figure 4.19 shows a photograph of the surface of the spalled Cu + 0.1% Si sample. The sample consisted of large grains up to 10 mm in size. It can be seen that the surface consists of dimples formed after the coalescence of the pores. Such surfaces are typical for ductile failure. Large grains are separated by shallow grooves. Details of the structure of intergranular fracture are visible in the micrographs shown in Fig. 4.20. The smaller size of the dimples

Fig. 4.19. A sample of Cu + 0.1% Si with a thickness of 3 mm after being hit by a copper plate 0.5 mm thick at a speed of 480 m/s.

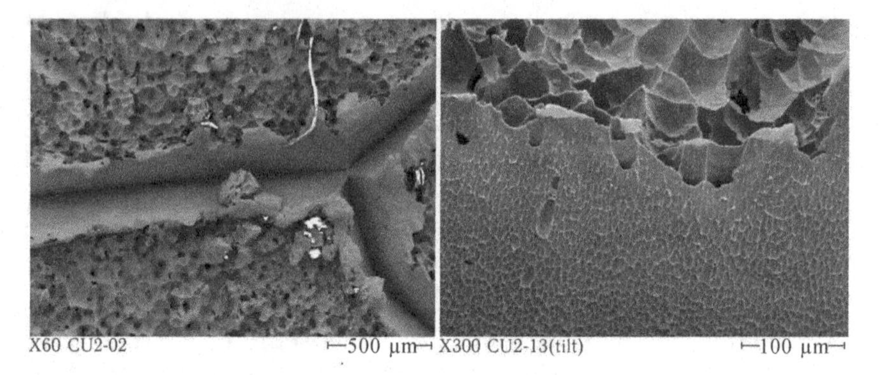

Fig. 4.20. The fracture surface of the Cu + 0.1% Si sample near the grain boundaries (scanning electron microscope).

Fig. 4.21. The spall fracture surfaces of Cu + 0.1% Si (left) and Cu + SiO_2 (right) samples.

at the grain boundaries means a higher concentration of the nucleation centres of fracture, which explains the reduced strength of polycrystalline copper.

Figure 4.21 shows a comparison of the spall fracture surfaces of Cu + 0.1% Si and Cu + SiO_2 samples. In the case of a solid solution of Cu + 0.1% Si, the dimensions of the dimples vary widely; there are two main populations with a size of ~4 microns and ~40 microns.

In the case of copper with the submicron silicon oxide inclusions, the dimples are very uniform and have a size of about 2 μm, corresponding to the average distance between the inclusions.

The larger size of the fracture dimples in the Cu + 0.1% Si sample means a longer pore growth time before their coalescence, which explains the prolonged nature of the spall fracture in this material. Comparison of the wave profiles in Fig. 4.18 and photographs of the fracture surfaces suggest that the uneven distribution of silicon in the Cu + 0.1% Si solid solution causes heterogeneous nucleation of voids in addition to the homogeneous process in the single crystal.

Thus, the experimental data definitely demonstrate the influence of structural factors on the resistance to high-rate fracture. It seems useful to formalize the results of observations by introducing an idea of the spectrum of material defects – potential foci of fracture characterized by different levels of stress required for their activation.

4.1.5. Spall strength of single crystals and polycrystals near melting

Figures 4.22 and 4.23 summarize the results of the first measurements of the spall strength of metals in polycrystalline and single-crystal states at temperatures up to the melting point. The thermodynamic estimates of the damage thresholds associated with the onset of melting under tension are also given there. The data show that polycrystalline aluminium and magnesium lose strength as they approach the melting point, while single crystals retain high tensile strength even after crossing the melting phase boundary in the negative pressure domain.

It can be assumed that the measured values of the dynamic strength of single crystals at high temperatures characterize the properties of partially molten aluminium. However, this assumption does not explain the differences in the behaviour of single crystals

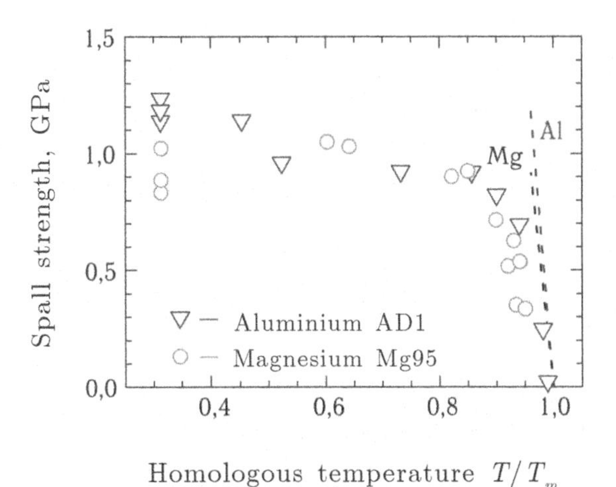

Fig. 4.22. The dependence of the spall strength on temperature for polycrystalline aluminium and magnesium [37] of commercial purity.

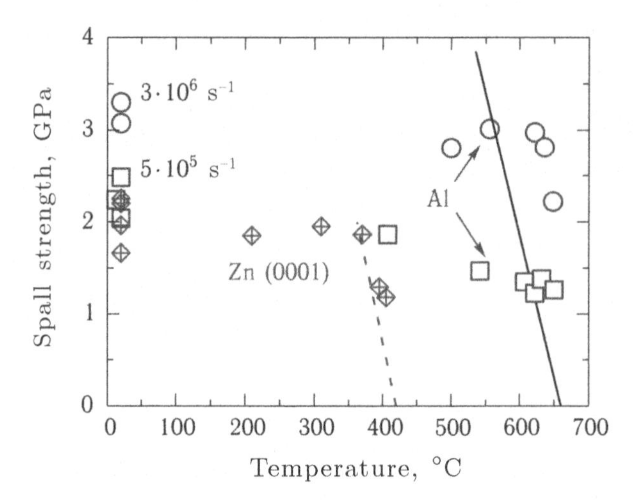

Fig. 4.23. Dependence of the spall strength on temperature for single crystals of zinc [38] and aluminium [17]. Lines show estimates made on the assumption of fracture at the onset of melting under tension. The values of the strain rate in the experiments with aluminium single crystals are indicated.

and polycrystalline aluminium. If melt zones appear inside the crystal, the crystal ceases to be homogeneous, and its strength properties should become the same as that of the polycrystalline material. However, even at the highest temperatures, single crystals of aluminium exhibit higher strength than polycrystalline aluminium at room temperature and the same strain rate. It seems more probable that the single crystal material did not melt in the experiments performed, and the measured strength in all cases corresponds to the strength of the solid. It is also interesting to note that the density of aluminium under maximum tension in the experiments performed has never decreased to the density of the aluminium melt at zero pressure.

If the expected melting during high-rate tension did not occur at high temperatures, then, in experiments with single crystals, states of a superheated solid were realized. The magnitude of the overheating reached 60–65°C with the shortest duration of the shock load. It is believed that the crystal surface plays a critical role in melting, where the activation energy is close to zero. The melting of a uniformly heated solid always begins on its surface. Overheated solid states can be created only inside the body, provided that its surface has a temperature below the melting point. This condition was implemented in the experiments. Once the melt nuclei have originated in a superheated solid, they should quickly grow, causing plastic deformation to form a surrounding crystalline material and provoking its fracture.

On the other hand, in polycrystalline materials it is possible to melt at the grain boundaries at temperatures below the melting point of the crystal [39]. Molecular dynamics calculations [40] show that melting in the bulk of the polycrystals begins at grain boundaries at temperatures that are noticeably lower than the melting temperature. In terms of thermodynamics, this means excessive internal energy of the surface layers of the grains, where the crystal structure is distorted due to the violation of the symmetry of the acting forces. It can be hoped that measurements of the dynamic strength of materials near the melting point will make it possible to estimate quantitatively the energy of the surface layers of the grains.

In the first measurements of the high-temperature spall strength of the metals shown in Fig. 4.22, there was a drop in the resistance to spalling fracture of the commercial aluminium AD1 and cast magnesium Mg95 to almost zero with approaching the melting point, which is explained by premelting in the negative pressure domain. Later experiments were conducted with high-purity metals and alloys, the results of which contradict this explanation. In high-purity polycrystalline metals – aluminium, copper, silver, premelting was not fixed, and aluminium alloy 6061 [41] retained some, albeit lowered, values of the dynamic elastic limit and spall strength even when the melting start temperature (solidus temperature) was exceeded.

In order to find out whether this difference is due to different material properties or a systematic experimental error in various measurements, comparative experiments were carried out under the same conditions with commercial aluminium AD1 and aluminium A999 with a purity of 99.999% at temperatures up to 640°C (20 degrees below the melting point). Experiments confirmed a sharp drop in the spall strength of commercial aluminium near the melting point, which is probably due to the early onset of melting at the grain boundaries where impurities are concentrated. Aluminium A999 retains a high spall resistance near the melting point.

5

Specific features of high-rate deformation of metals in shock compression

5.1. General views

Investigations of the temperature and rate dependences of the resistance to deformation of metals and alloys allow us to study the basic laws of movement of carriers of plastic deformation – dislocations, to identify the governing factors and laws of the formation and development of damage in a material. This information is needed to optimize the machining of materials, as well as to solve problems of high-speed impact and penetration. This section presents the most exotic research results of the last decade. To discuss them, it is useful to recall some basics of modern ideas about the mechanisms of high-rate deformation and fracture of solids.

In terms of the theory of dislocations, the rate of plastic deformation $\dot{\gamma}$ is determined by the average velocity of the mobile dislocations v_d and their density N_m, connected by the Orowan relation

$$\dot{\gamma} = bN_m v_d, \tag{5.1}$$

where b is the Burgers vector. The average velocity of mobile dislocations is a function of stress, temperature, and the concentration of various kinds of defects that impede the movement of dislocations, including the dislocations themselves. Of course, in addition to dislocations, twinning can make a significant contribution to the mechanism of plastic deformation, which is especially significant for crystals with the HCP and BCC structures. This, however, in most cases does not introduce radical changes in the interpretation of the rate and temperature dependences of the flow stress during high-rate deformation of metals. The limited integral information obtained in experiments on the laws of high-rate deformation, as a rule, also does not allow one to identify the contributions of various types of dislocations. For these reasons, in the following we will use the dislocation terminology in a certain averaged and simplified sense, without going into details of the mechanism of high-rate deformation inaccessible for modern experiments.

It is known that the flow stress of crystalline solids increases with increasing loading rate. For many metals, this dependence sharply increases with an excess of the strain rate of $\sim 10^3$–10^4 s^{-1}, which is interpreted as a consequence of a change in the mechanism of motion of dislocations. At low strain rates, the dislocations overcome the Peierls barriers and obstacles as a result of the joint action of the applied stress and thermal fluctuations. As a consequence, an increase in temperature is accompanied by a decrease in the yield strength of materials. For deformation at high rate, it is not necessary to apply higher stresses. At a sufficiently high strain rate, the effective stresses are so high that dislocations are able to overcome barriers and obstacles without the additional contribution of thermal fluctuations. In this case, the dominant drag mechanism becomes phonon viscosity – the force of resistance caused by thermal vibrations of atoms. Since the phonon viscosity is proportional to temperature, at very high strain rates one can expect a linear increase in the flow stress with increasing temperature [42]. At a sufficiently large stress, called 'ideal' or 'ultimate' shear strength, the material must lose stability with respect to shear stresses and may deform without any contribution from dislocations. The value of τ_{id} of the ideal shear strength is proportional to the shear modulus G and, according to various estimates, is $\tau_{id} \approx G/10...G/2\pi$. Since the shear modulus decreases with temperature, the ideal shear strength should also decrease when heated. The ratio of the contributions of the thermal-fluctuation and over-barrier mechanisms of dislocation motion as a function of temperature and strain rate is illustrated in Fig. 5.1.

The second equally important parameter determining the resistance to plastic deformation is the total dislocation density. Figure 5.2 explains the dependence of the yield strength on the density of

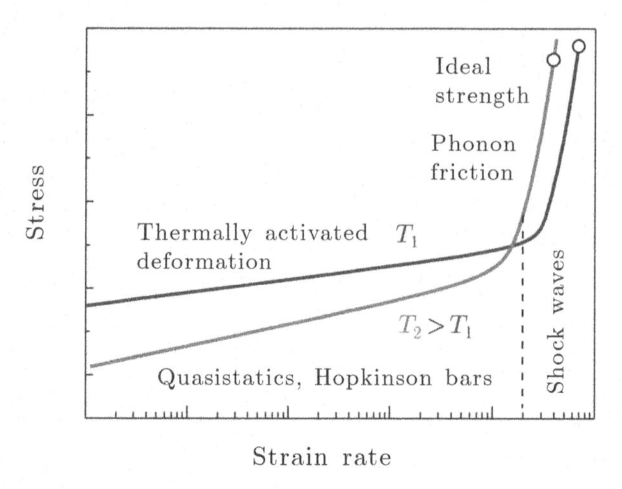

Fig. 5.1. The explanation of the mechanisms of temperature–time dependences of yield strength. Specified research methods.

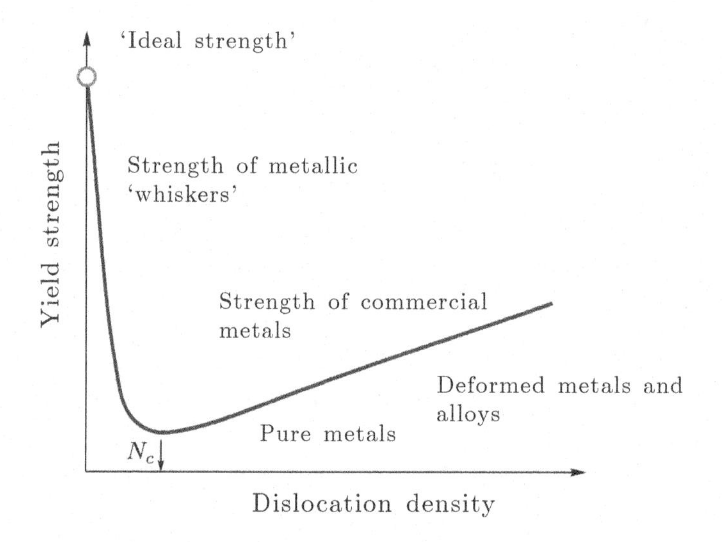

Fig. 5.2. A schematic representation of the dependence of the stress of plastic flow of a crystalline body on the dislocation density.

dislocations. Defect-free crystals are characterized by the highest yield strength values. High-strength structural states similar or close to them are realized in metallic 'whiskers' of micron thickness. With the appearance of dislocations, the flow stress rapidly decreases to a minimum at the critical density N_c, and with a further increase in the density, the dislocations begin to block each other, as a result of which the flow stress increases. An increase in the density of mobile dislocations occurs in the process of plastic deformation.

In the experiments under discussion, one way or another, a plane shock wave is created in the plate of the material under study and its structure is measured – as a rule, by recording the velocity of the free surface as a function of time. The thickness of the plane specimen which determines the propagation time of a shock wave can vary from 50–100 μm to 10 mm or more, and the velocity profile of the free surface using modern experimental techniques is measured with a resolution of 1 ns (10^{-9} s). Most recently, it has become possible to experiment with samples of micron and submicron thickness, while the temporal resolution of the measurements reaches the picosecond (10^{-12} s) level.

Figure 5.3 shows, for example, the free surface velocity histories $u_{fs}(t)$ of flat specimens of magnesium alloy Ma2-1 of different thickness measured in experiments at room temperature. The shock compression pulse was generated in the samples by the impact of a plate whose thickness was several times smaller than the thickness of the sample.

Due to a sharp increase in compressibility during the transition from elastic uniaxial compression to plastic, the shock wave loses stability and splits into an elastic precursor, which propagates at a speed close to the longitudinal sound velocity c_l, and the plastic wave behind the speed of which is determined by the bulk compressibility of the material. The compression stress in the elastic precursor is equal to the Hugoniot elastic limit (Hugoniot Elastic Limit, HEL). According to the measured profile of the free-surface velocity, the value of the Hugoniot elastic limit is defined as

$$\sigma_{\mathrm{HEL}} = \rho c_l u_{\mathrm{HEL}}/2, \tag{5.2}$$

where u_{HEL} is the velocity of the free surface behind the front of the elastic precursor, ρ is the density of the material. Although the term 'Hugoniot elastic limit' is generally accepted, it is, strictly speaking, not quite correct, since the value of σ_{HEL} is not a constant. From the comparison of the wave profiles in Fig. 5.3 it can be seen that, despite the growing parameters behind the front of the elastic precursor on each wave profile, the stress at the front of the elastic precursor decreases as the wave propagates. This decay of the elastic precursor is a consequence of the relaxation of stresses in the process of plastic deformation directly behind the elastic shock wave.

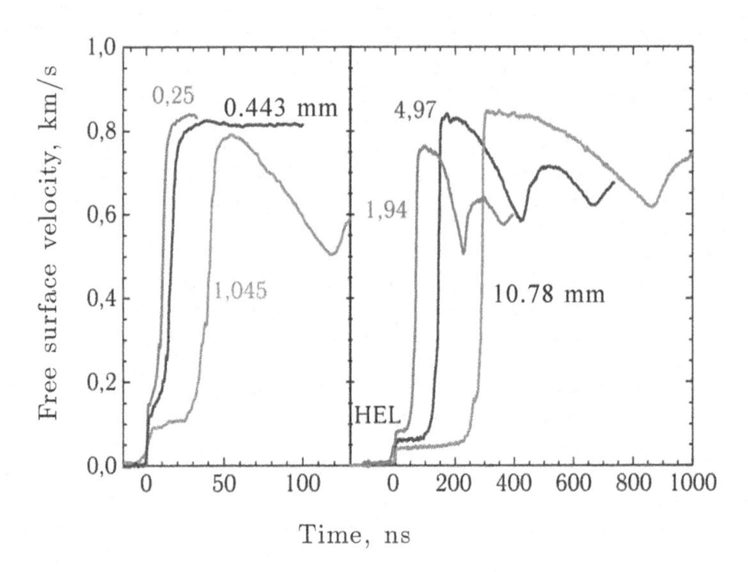

Fig. 5.3. The results of measurements of the free surface velocity histories of samples of magnesium alloy Ma2-1. Numbers at the wave profiles indicate the sample thickness.

5.2. Shock wave investigation methodology of rate dependences of deformation resistance

There are two ways to directly obtain information on the relationship between the rate of plastic deformation and flow stress, based on measurements of the decay of the elastic precursor [9, 43] and measurements of the width of the plastic shock wave [10, 44]. These research methods will be briefly described below, after which the most interesting results will be presented. More information about the macrokinetic patterns of high-rate deformation and fracture is obtained using the methods of computer simulation of shock-wave experiments with one or another hypothetical models and constitutive relations. This direction is not considered here.

The decay of the elastic precursor of the shock compression wave due to the relaxation of stresses is associated with the rate of plastic deformation behind its front $\dot{\gamma}=\left(\dot{\varepsilon}_x^p-\dot{\varepsilon}_y^p\right)/2$ by the relation (3.31), which we rewrite here for convenience in the form:

$$\left.\frac{d\sigma_x}{dh}\right|_{\text{HEL}}=-\frac{4}{3}\frac{G\dot{\gamma}_p}{c_l}, \tag{5.3}$$

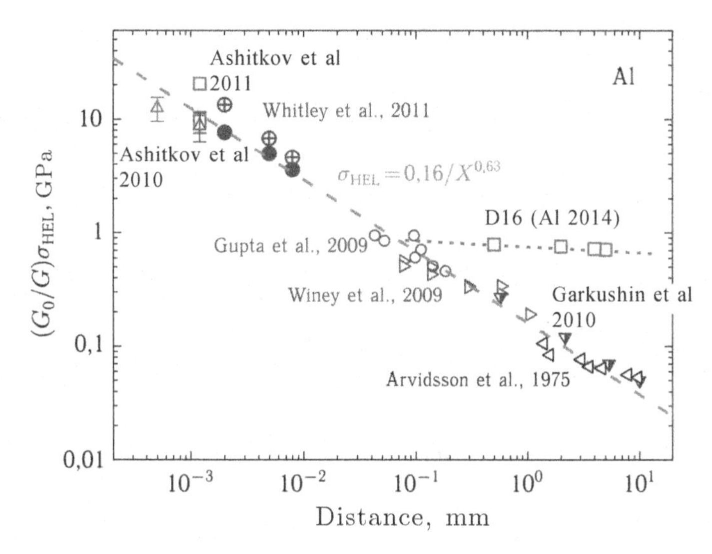

Fig. 5.4. The dependence of the amplitude of the elastic shock wave in aluminium on the distance travelled. For comparison, the data for D16 aluminium alloy are also given.

where h is the distance travelled by the wave, G is the shear modulus, c_l is the velocity of propagation of the precursor front, adopted in this approximation equal to the longitudinal velocity of sound. Figure 5.4 summarizes experimental data on the decay of an elastic precursor in aluminium, including data in the picosecond range of durations obtained on film samples with a thickness of about 1 μm. The compressive stress behind the precursor front, equal to the value of the Hugoniot elastic limit (HEL) of aluminium, according to measurements, varies from 50 MPa at a distance of 10 mm to 20.5 GPa at a distance of 1.2 μm. Note that elastic compression up to 20.5 GPa is realized also in the middle of a steady plastic shock wave having the same propagation velocity U_S = 7.8 km/s, with a final shock compression pressure of 38.7 GPa. Such a strong compression leads to a significant increase in the shear modulus in the relation (5.3). The solid symbols in Fig. 5.4 show the normalized values of $\sigma_{HEL} G_0/G$. With this correction, the entire set of experimental data with reasonable accuracy is written by an empirical relation

$$(G_0/G)\,\sigma_{HEL} = S(h/h_0)^{-\alpha}, \qquad (5.4)$$

where h_0 = 1 mm, S = 0.16 GPa, exponent α = 0.63. At compressive stresses less than 1 GPa, which in Fig. 5.4 corresponds to distances greater than several tens of micrometers, the increase in the shear modulus is insignificant and we will neglect this in the future. The magnitude of the maximum shear stress behind the precursor front is equal to

$$\tau_{\text{HEL}} = (3/4)\sigma_{\text{HEL}}\left(1 - c_b^2/c_l^2\right) = \sigma_{\text{HEL}} G/E', \tag{5.5}$$

where is $E' = \rho_0 c_l^2$ is the modulus of longitudinal elasticity. As a result, the empirical dependence (5.4), taking into account (5.3) and (5.5), is transformed into the dependence of the initial rate of plastic deformation on the shear stress:

$$\dot{\gamma}_p = \frac{3}{4}\left(\frac{\tau E'}{SG}\right)^{\frac{\alpha+1}{\alpha}} \frac{S\alpha c_l}{h_0 G} \quad \text{or} \quad \dot{\gamma}_p = 9.1 \cdot 10^7 \left(\tau/\tau_0\right)^{2.59} \text{s}^{-1}, \tag{5.6}$$

where $\tau_0 = 1$ GPa. As can be seen from the graphs in Figs. 5.5 and 5.6, the initial plastic strain rate behind the front of the elastic precursor decreases as it propagates from 10^9 s^{-1} at a distance of 1 μm to 10^3 s^{-1} by 5–10 mm, and further the decay of the precursor slows down sharply. We note that experiments with the Hopkinson bar [45] demonstrate a sharp increase in the flow stress at a strain rate of ~(2–5) · 10^3 s^{-1}.

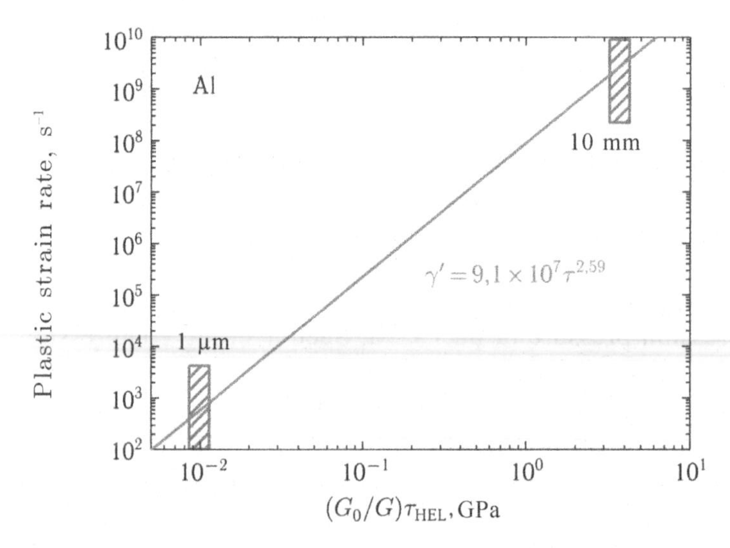

Fig. 5.5. The dependence of the initial rate of plastic deformation behind the front of the elastic precursor in aluminium on the magnitude of the shear stress, calculated in accordance with the relation (5.6) according to the graph in Fig. 5.4.

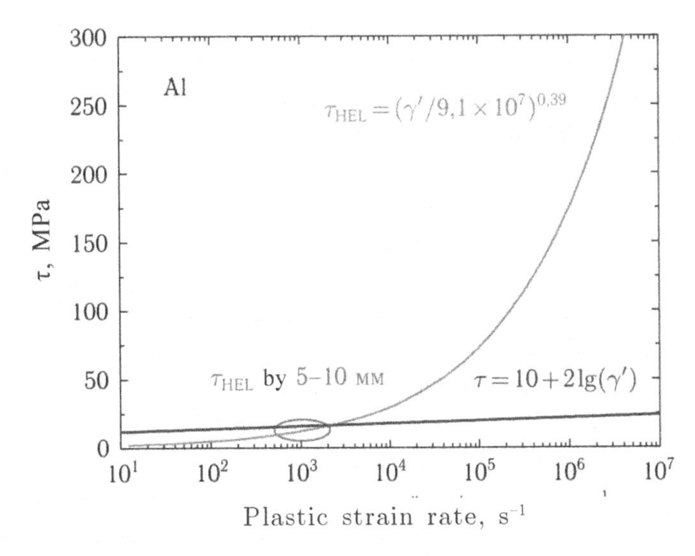

Fig. 5.6. Aluminium flow stress at moderate (logarithmic function) and high strain rates.

Figure 5.4 also presents the data on the attenuation of the elastic precursor in the aluminium alloy D16. A stronger alloy demonstrates a very weak decay of the elastic precursor at its propagation from 0.5 mm to 5 mm. In this range of distances travelled, the Hugoniot elastic limit of the alloy is approximately 0.71 GPa. In aluminium, such a stress value behind the front of an elastic precursor is realized at a distance of 92 μm, where the plastic strain rate is $\sim 7 \cdot 10^5$ s^{-1}. It is natural to assume that for samples of smaller thickness, the behavior of the alloy should not differ so much from aluminium.

The second source of data on the behavior of materials at extremely high strain rates is the compression rate in a plastic shock wave. In principle, the total strain rate $\dot{\varepsilon}_x$ in a stationary shock wave is determined quite simply – by differentiating the corresponding section of the particle velocity history $u_p(t)$ and dividing by the wave propagation velocity U_S: $\dot{\varepsilon}_x = \dot{u}_p / U_S$. On the free surface velocity history $u_{fs}(t)$, the recorded rise time in a plastic shock wave can be slightly overestimated, and the apparent strain rate, respectively, is underestimated as a result of multiple reflections of the elastic wave between the free surface and the plastic front.

The determination of shear stresses in the shock wave is not so clear. Approximation of a steady wave assumes the constancy of its form and the constancy of the parameters of the state of matter before and after it, which, generally speaking, does not fully correspond

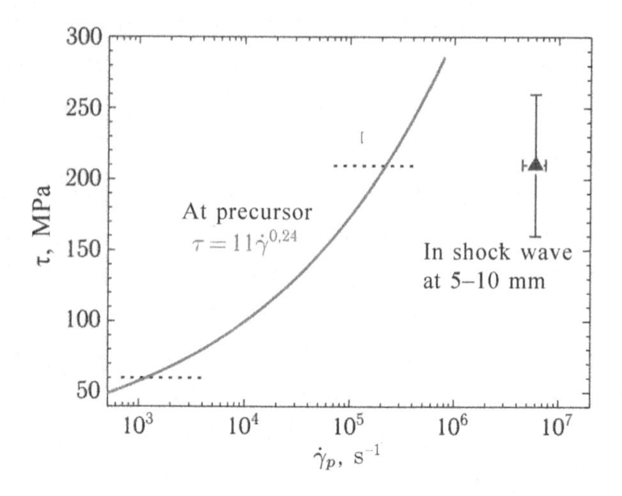

Fig. 5.7. The ratio between the plastic strain rate $\dot{\gamma}_p$ and the shear stress τ behind the precursor front (solid line) and in the plastic shock wave (triangle) according to the results of experiments with the magnesium alloy Ma2-1. The dotted lines mark the experimental range of the parameters of the elastic precursor.

to the experimental data. For this reason, the data discussed below should be considered as estimated.

The total strain rate $\dot{\varepsilon}_x$ is defined for the point of maximum compression rate on the profile $u_{fs}(t)$ as $\dot{\varepsilon}_x = \dot{u}_{fs}/2U_S$. The maximum shear strain rate under uniaxial compression is $\dot{\gamma} = \dot{\varepsilon}_x/2$, is the sum of the elastic component $\dot{\gamma}_e = \dot{\tau}/2G$, and the plastic shear strain rate $\dot{\gamma}_p$. As a result, we have:

$$\dot{\gamma}_p = \frac{\dot{\varepsilon}_x}{2} - \frac{\dot{\tau}}{2G}. \tag{5.7}$$

In a steady plane wave, a change in the state of matter occurs along the Rayleigh line, which is the straight line $\sigma_x = -\rho^2{}_0 U^2{}_S (V - V_0)$ connecting the states before the wave and behind it. The deviator stress component in the wave is the difference between the stress σ_x on the Rayleigh line and the pressure p on the Hugoniot of a substance at the same degree of compression. In this case, the shear stress $\tau = (3/4)\,(\sigma_x - p)$ as compression progresses through a maximum at some intermediate point. At the maximum point, $\dot{\tau} = 0$ and $\dot{\gamma}_p = \dot{\varepsilon}_x/2$. The corresponding magnitude of the shear stress is estimated as its value at the maximum point plus the magnitude of the shear stress in front of the wave, which was determined in the same way as τ_{HEL}, but using a higher surface velocity in the middle of the wave profile between the precursor front and the plastic shock wave.

The values of the shear stress τ and the strain rate found in this way from the results of experiments with the magnesium alloy Ma2-1 are shown by the dots in Fig. 5.7. The pressure of shock compression in these experiments was 3.8 GPa. The error in determining τ is related mainly with the lack of information on the stress state of the alloy in a shock-compressed state. Figure 5.5 shows that the rate of plastic deformation in a shock wave is an order of magnitude greater than that in an elastic precursor with the same value of shear stress. A similar ratio of strain rates was observed in experiments with most other metals and alloys with different crystal structures. In all likelihood, such a decrease in the characteristic viscosity $\tau/\dot{\gamma}_p$ of the material with the development of plastic deformation is explained by the intense multiplication of dislocations. In this regard, it is appropriate to discuss the Swegle–Grady empirical relation [10], which is still intriguing for many researchers. After analyzing the results of measurements of the rise time of a plastic shock wave for different materials, Swegle and Grady found that all of them can be described with acceptable accuracy by power law functions of the final shock compression pressure with the same exponent equal to 4.

Figure 5.8 compares the results of measurements of the compression rate in the shock wave for titanium and glycerol, which show that its dependence on the pressure of shock compression

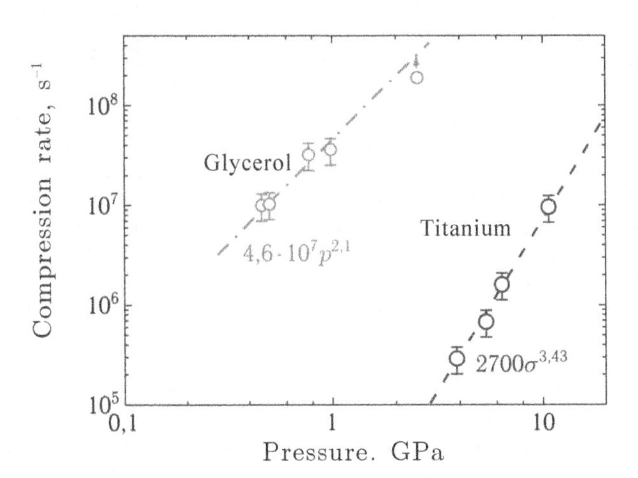

Fig. 5.8. The compression rate in the shock wave as a function of the shock compression pressure for titanium [46] and glycerol [47]. Numerical expressions of dependences are shown.

is much weaker for a liquid than for a solid. The reason for the difference is related to the different physical nature of the viscosity of liquids and solids: the viscosity of liquids is known to be determined by the momentum transfer during the chaotic movement of molecules, while the viscosity of crystalline solids is determined by the dislocation dynamics. In particular, the strong dependence of the compression rate in a shock wave on its intensity is apparently the result of rapid multiplication of dislocations in the process of high-speed plastic deformation.

5.3. Temperature effects

The effect of temperature and strain rate on the dynamic yield strength under shock-wave loading of relatively high-strength metals and alloys is not different or not very different from that expected based on extrapolation of data obtained at lower strain rates. For example, Figure 5.9 summarizes the data on the yield strength values of high-strength titanium alloy Ti-6-22-22S [48] at various strain rates obtained by different methods. The results of shock-wave measurements, measurements by the Hopkinson bar method at moderate strain rates and standard measurements at low-rate deformation, in general, are in complete agreement and are described by a single logarithmic dependence. For the model illustrated in fig. 5.1, this means the preservation of the thermoactivation mechanism

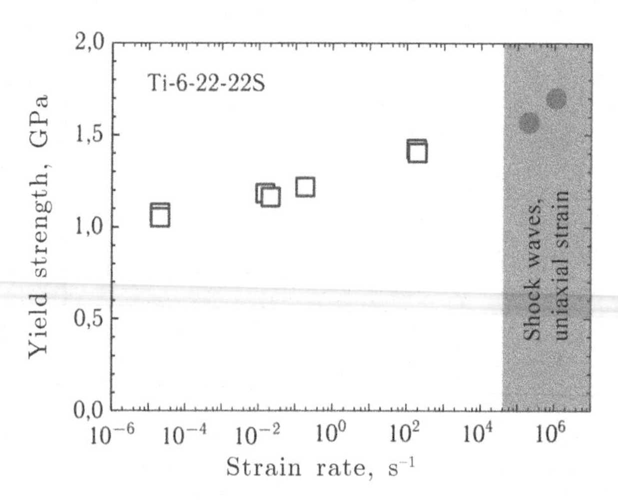

Fig. 5.9. The dependence of the yield strength of high-strength titanium alloy Ti-6-22-22S according to measurement results by standard methods at low-rate deformation, on Hopkinson bars at moderately high strain rates and shock-wave measurements.

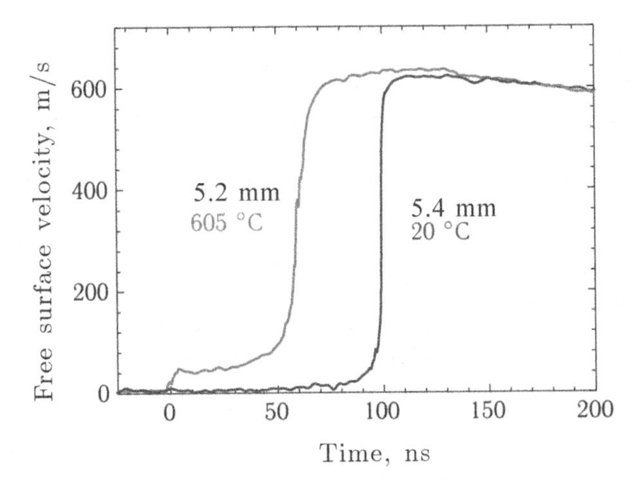

Fig. 5.10. The initial sections of the wave profiles of shock compression of aluminium AD1 samples with a thickness of 5 mm at temperatures of 20°C and 605°C [24].

of deformation corresponding to the low-rate branch of the general dependence. With increasing temperature, the dynamic yield strength of this alloy decreases.

For a number of metals and ionic crystals, an anomalous increase in the Hugoniot elastic limit with increasing temperature was experimentally observed. The effect is illustrated in Fig. 5.10, where the free surface velocity histories of technical aluminium samples measured at normal and elevated temperatures are compared. An increase in temperature led to a significant increase in the amplitude of the elastic precursor and an increase in the rise time of parameters in the plastic shock wave: from 3–5 ns at room temperature to 8–12 ns at 605°C. Note that the increase in the precursor amplitude is partly due to a decrease in the longitudinal sound velocity and, accordingly, an increase in the Poisson ratio. The decrease in the longitudinal sound speed with heating is manifested in the wave profile also in the decrease in the time interval between the precursor front and the plastic shock wave. However, the effect does not boil down to a reduction in the impedance of the material; the calculation of shear stresses from σ_{HEL} using the values of elastic moduli at a given temperature confirms the anomalous increase in the flow stress under these conditions.

Figure 5.11 compares [49] the results of measurements of the effect of temperature on the dynamic yield strength of titanium and its alloys. It is seen that the anomalous increase in the dynamic yield

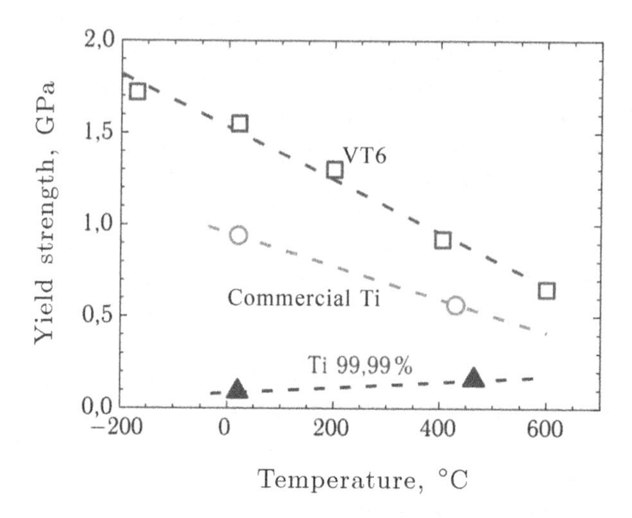

Fig. 5.11. Dependences of the dynamic yield strength of titanium and its alloys on temperature as determined from shock wave measurements.

strength during heating occurs for pure titanium, where the stress of the high-rate plastic flow is comparable with the forces of phonon viscosity. The yield strength of the alloys is increased by artificially created obstacles to the movement of dislocations and significantly exceeds the dislocation dragging stress by phonon viscosity; as a result, anomalous thermal hardening is not recorded for the alloys.

5.4. Approaching ultimate (ideal) shear strength

Let us consider, by the example of iron, the specific features of the behaviour of metals under ultrashort loading. Figure 5.12 shows the free-surface velocity histories of iron films with a thickness of 250 ±5 nm and 540 ± 5 nm, measured with laser pulses with a duration τ = 150 fs and an energy density at the centre of the focal spot of 3 J/cm^2 [50]. In experiments using an interferometric method with frequency modulated diagnostics, displacement of the free rear surface of the sample was recorded as a function of time with an error of 1 nm in the distance and 1 ps in time. The surface velocity histories are then obtained after several iterations, during which the best agreement between the velocity integral and the measured surface displacement history was achieved. The total speed error is estimated at about 10%.

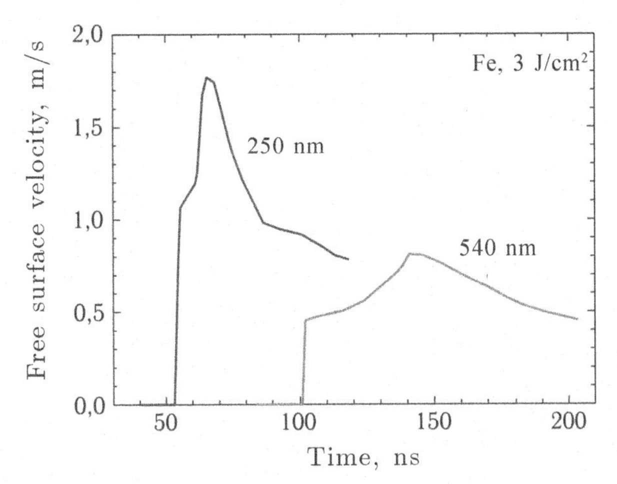

Fig. 5.12. The evolution of a compression pulse generated in iron film samples by a laser pulse of the femtosecond range.

The speed of propagation of the wave configuration front in the path from 250 nm to 540 nm on average over the entire amount of the experiments performed was U_S = 6.45±0.2 km/s. At the same time, the surface velocity behind the first shock wave decreases with propagation from 1.06 ± 0.06 km/s at a distance of 250 nm to 0.45 ± 0.03 km/s at a distance of 540 nm. The high wave propagation velocity and the short rise time of the parameters indicate that the first wave in the two-wave configuration is an elastic precursor. The compression stress behind the precursor front was 27.5±2.5 GPa at a distance of 250 nm and 11.0±1 GPa at 540 nm. Note for comparison that the Hugoniot elastic limit of sapphire in the sub-microsecond time range varies from 13 GPa to 24 GPa depending on the direction of compression, and the Hugoniot elastic limit of diamond is 50–100 GPa.

With such large stresses of elastic uniaxial compression, we can no longer limit ourselves to the approximation of the constancy of elastic moduli and sound speeds. For analysis and computer simulation, it would be convenient to have a description of large elastic compressions based on some simple assumption and with reasonable accuracy corresponding to the experimental data. It was previously shown that the natural approximation for estimating the longitudinal speed of sound on the basis of the assumption that the Poisson coefficient is constant agrees well with the available experimental data for metals in a wide range of shock compression

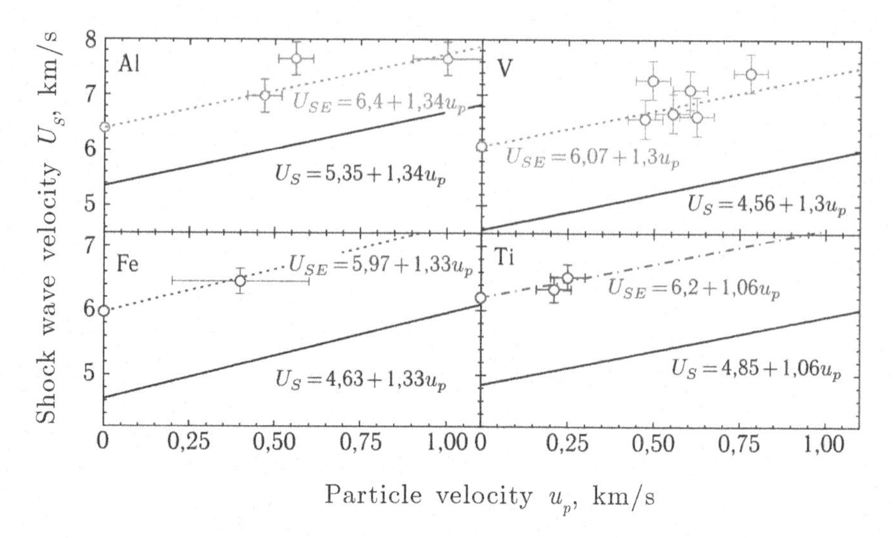

Fig. 5.13. The results of measurements of the propagation speeds of the US and particle velocities up behind the front of the elastic precursor in thin films of aluminium, iron, vanadium in comparison with the equilibrium adiabats and metastable Hugoniots of elastic compression, calculated under the assumption that the Poisson coefficient is constant.

pressures. The constancy of the Poisson's ratio means the constancy of the ratio of the longitudinal and bulk speeds of sound c_l/c_b. Using the quasi-acoustic approximation for the bulk speed of sound c_b, it can be shown that the ratio $c_l(V)/c_b(V)$ turns out to be the same, both on the adiabat of hydrostatic compression, and on the metastable adiabat of uniaxial compression only if coefficient b in a linear relationship $U_S = c_0 + bu_p$ between the velocity of the shock wave *US* and the particle velocity behind it up has the same value for both Hugoniots. The applicability of this approximation is illustrated in Fig. 5.13 on the example of four studied metals.

Figure 5.14 shows the corresponding state diagram of iron implemented in the picosecond range of load durations, which shows the equilibrium Hugoniot of iron with a transition to the ε-high-pressure ε-phase, the metastable Hugoniot $p(V)$ of the low-pressure α-phase and the metastable Hugoniot of uniaxial elastic compression. From the deviation of the state behind the precursor front from the equilibrium Hugoniot of the low-pressure phase, the magnitude of the maximum shear stress τ was determined from the relation

$$\sigma_z(V) - p(V) = \frac{4}{3}\tau,$$

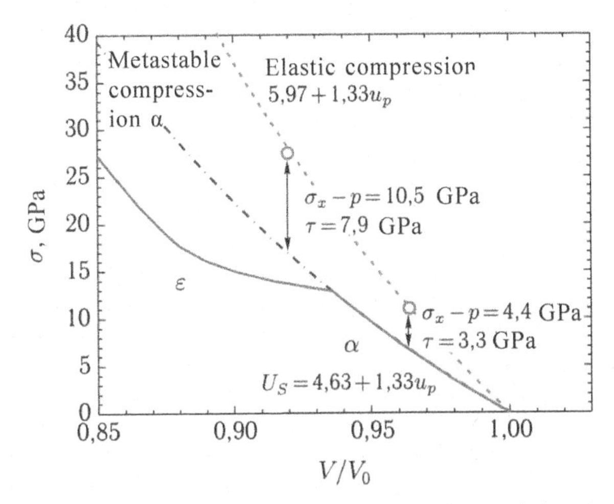

Fig. 5.14. Parameters of the state of iron behind the front of the elastic precursor at distances of 250 nm and 540 nm.

where σ_z is the longitudinal elastic compression stress. The obtained values of τ were 7.9 GPa and 3.3 GPa. The ultimate values of shear resistance – ideal shear strength, obtained from *ab initio* calculations, are 7.2–7.5 GPa for iron [59, 60]. The estimates based on the results of the measurements presented here gave a maximum value of τ = 7.9 GPa, that is, even slightly higher than the calculated value of the ideal shear strength. It should be noted, however, that under compression, both the shear modulus and, accordingly, the magnitude of the ideal shear strength, proportional to it, increase. In any case, it can definitely be argued that in the picosecond range of load durations, stress states of iron are realized and measured, very close to ideal strength.

Unfortunately, the considerable nonstationarity of the waves and the high rate of relaxation processes make it impossible to estimate the course of change in the state of the material after overcoming the Hugoniot elastic limit. In our experiments, we did not obtain convincing evidence of the α → ε polymorphic transformation in the picosecond time range. In [85], the results of measurements of the free-surface velocity histories of thicker iron samples (1.2–1.6 μm) indicate the possibility of polymorphic transformation in a time of the order of 100 picoseconds.

To date, measurements of the Hugoniot elastic limit and spall strength have been carried out near their maximum possible values for metals with the fcc, bcc, and hcp structures. Systematic measurements of the evolution of elastoplastic shock compression waves in the submicrosecond range of durations were carried out for these metals. All sets of experimental data, as a rule, are reasonably consistent and form the basis for the construction of wide-range models of high-rate deformation. Experimental data in the picosecond range are also needed for testing and setting problems for an atomistic modelling of the deformation and fracture processes.

5.5. Effects of annealing and hardening treatment

It is known that annealing of a metal after its rolling or another type of severe plastic deformation reduces the density of dislocations in it and, as a result, decreases the yield strength. This, however, is not always true for high-rate deformation under sub-microsecond shock-wave loading. Figure 5.15 demonstrates the difference in the effects of annealing of pure copper and tantalum. While the Hugoniot elastic limit of copper decreased many times after annealing, for tantalum and other metals with a BCC structure, annealing led to an increase in HEL. Figure 5.16 shows the annealing effect of commercial titanium having an HCP structure; in this case, not only the value of the Hugoniot elastic limit has changed, but also the shape of the elastic precursor on the wave profile. The formation of a peak in the frontal part is connected, in all likelihood, with the intensive multiplication of dislocations or their unblocking from the cloud

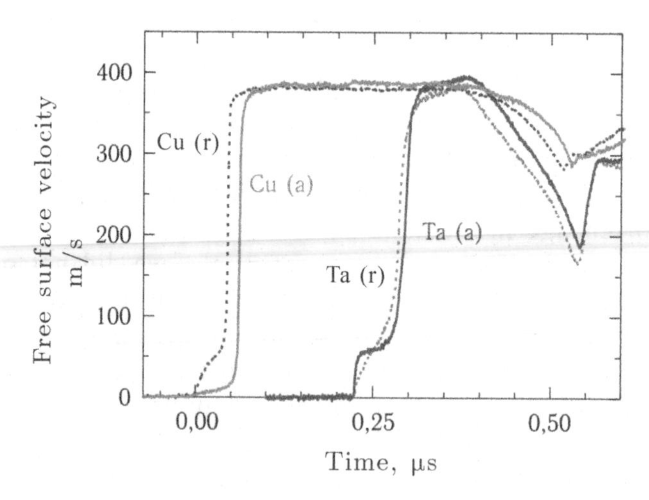

Fig. 5.15. Free surface velocity histories of 2 mm plates in the state after rolling (dotted line index *r*) and after annealing (solid lines, index *a*) for Cu and Ta.

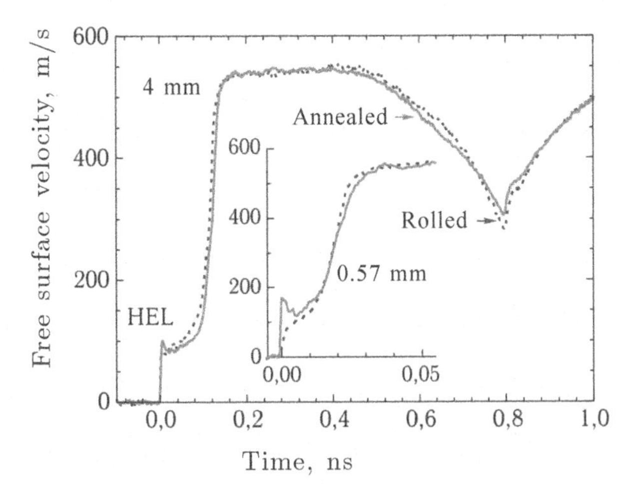

Fig. 5.16. The free surface velocity histories of as-rolled and annealed samples of commercial titanium VT1-0 with a thickness of 4 mm and 0.57 mm [64].

of impurities. In principle, this observation does not contradict the existing ideas about the correlation between the plastic flow stress and a dislocation density, discussed in section 5.1 (Fig. 5.2). We note once again that for all the metals under discussion, the hardness after annealing decreased. Consequently, the 'critical' dislocation density at which the flow stress takes on a minimum value is different for different strain rates.

The examples of the wave profiles show that after annealing, not only the Hugoniot elastic limit changed, but also the time interval between the elastic and plastic waves increased. This effect is shown in more detail in Fig. 5.17, where the results of measurements [51] of this interval for the rolled and annealed tantalum depending on the distance traveled by the wave are compared. At distances $h < 1$ mm, the recorded velocity of a plastic shock wave is equal to the bulk velocity c_b in tantalum, although according to the theory it must exceed it; in some other cases, the recorded velocity of the plastic shock wave may even be less than the speed of sound. Subsonic plastic shock wave velocity is often associated with loss of shear strength during shock compression. Such an interpretation, however, can be valid only for steady waves; the graph in Fig. 5.17 rather indicates the process of establishing the stationarity of elastic and plastic waves. The nonsteady processes in relaxing media require more detailed analysis which has no as yet been carried out.

Observations of the anomalous effect of annealing on the magnitude of the dynamic yield strength correlate with the results of

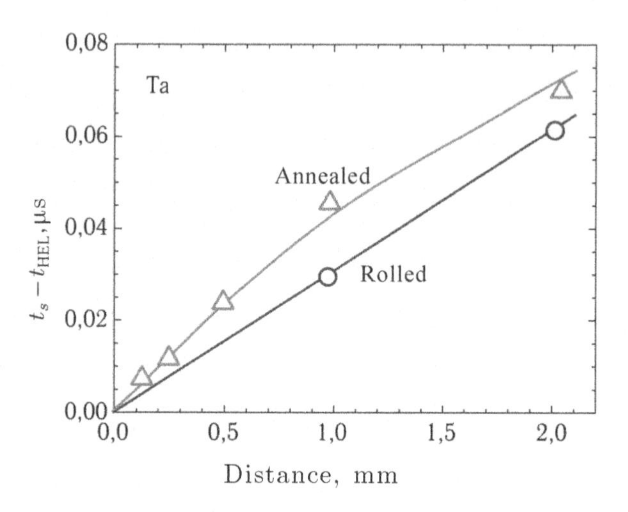

Fig. 5.17. The time intervals between the elastic and plastic waves in tantalum in the states after rolling and after annealing as a function of the distance travelled.

studies of the effects of refining the grain structure using the method of severe plastic deformation [52]. While such processing is always accompanied by a significant increase in the hardness of the material, the results of flow stresses measurements in the sub-microsecond range of load durations are not so unambiguous. In particular, the increase in the value of the dynamic yield strength of aluminium alloys with a decrease in the grain size roughly corresponds to low-rate test data; for titanium VT1-0 and tantalum instead of increasing there is a slight drop in the dynamic yield strength; for copper, iron, and titanium alloy VT6, the effect is substantially less than under conditions of low-rate deformation.

As an example Fig. 5.18 compares the results of measurements of the free surface velocity histories $u_{fs}(t)$ of samples of the initial coarse-grained (CG) and ultrafine grained (UFG) tantalum. Their hardness is 76–79 and 103–104 units of HRB, respectively, that is, severe plastic deformation led to a thirty percent increase in hardness. Nevertheless, the date shown in Fig. 5.18 show that the value of the Hugoniot elastic limit σ_{HEL} after refining the grain structure turned out to be even slightly lower than for the original coarse-grained material. This difference is especially evident in experiments with samples with a thickness of ~0.65 mm. The effect is explained by a higher stress relaxation rate in the fine-grained material. Accelerated stress relaxation in the fine-grained material is also manifested in a higher compression rate in the plastic shock wave.

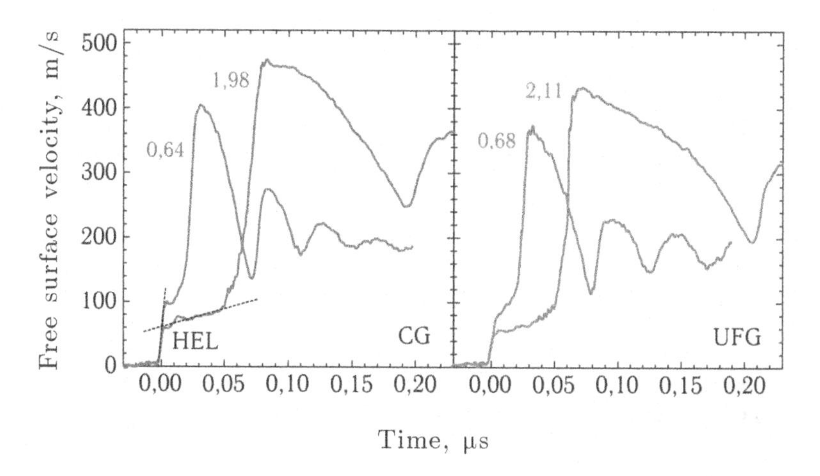

Fig. 5.18. The results of measurements of the free surface velocity histories of the samples of the initial (CG) and ultrafine grained (UFG) tantalum. Numbers in the histories indicate the thickness of the samples.

The results of the studies clearly demonstrate the anomalous effect of hardening treatment on the resistance to high-rate deformation and fracture. Phenomenologically, a decrease in the dynamic yield strength in compression after rolling could be identified as a manifestation of the Bauschinger effect, according to which a certain proportion of plastic deformation is reversible and does not contribute to strain hardening. The Bauschinger effect is also favoured by the numerous results of the measurements of full stress pulses, which show that the deformation process during unloading after shock compression deviates from the elastoplastic towards smaller deviatoric stresses, and plastic deformation very often begins directly behind the unloading wave front without a finite elastic section. A different interpretation is based on the fact that imperfections of the structure can serve not only as obstacles for the movement of dislocations and twins, but at high stresses, as their sources, thereby weakening the dependence of the yield strength on the strain rate.

To understand the results of the experiments performed, it is obviously necessary to take into account that under the action of applied shear stresses not only displacement of dislocations occurs, but also their nucleation and multiplication. The origin of dislocations occurs near stress concentrators, which are violations of the crystalline long-range order. The imperfections of the structure, on the one hand, are a strengthening factor, and on the other hand, they represent sources of carriers of plastic deformation (dislocations). In other words, the same defects can determine an increased resistance to deformation under quasistatic conditions and can be sources of dislocations at high rates of deformation and, accordingly, high stresses and thus can reduce the resistance to plastic deformation. It follows from the presented data that the difference in the rate dependences can be so large that with the transition from quasi-static to high-rate loading, the influence of these defects on the flow stress can change sign.

5.6. Temperature–rate dependences of the flow stress at different stages of shock-wave deformation of metals

Figure 5.19 presents experimental data [53] on the temperature–rate dependence of the flow stress of high-purity aluminium in the range of strain rates of 10^4–10^6 s^{-1}, obtained from measurements of the decay of an elastic precursor and the width of a plastic shock wave at temperatures up to the melting point. It can be seen, in particular, that at the same shear stress, the rate of plastic deformation increases by at least an order of magnitude after deformation in a shock wave of ~2%. The shear stress in the elastic precursor τ_{HEL} is related to the rate of plastic deformation $\dot{\gamma}$ by the relationship

$$\tau_{HEL} = A\dot{\gamma}^{\frac{\alpha}{\alpha+1}}, \tag{5.8}$$

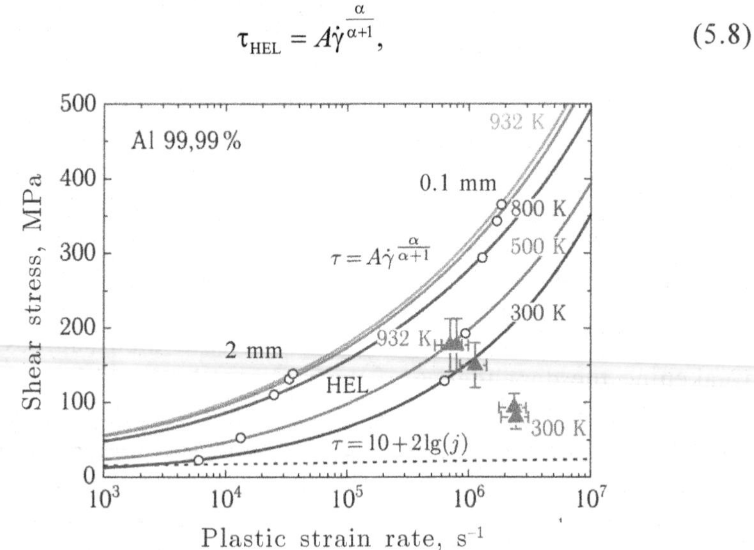

Fig. 5.19. Dependence of the initial flow stress (lines) and flow stress after 2% deformation (points) on the plastic strain rate and temperature for high-purity aluminium.

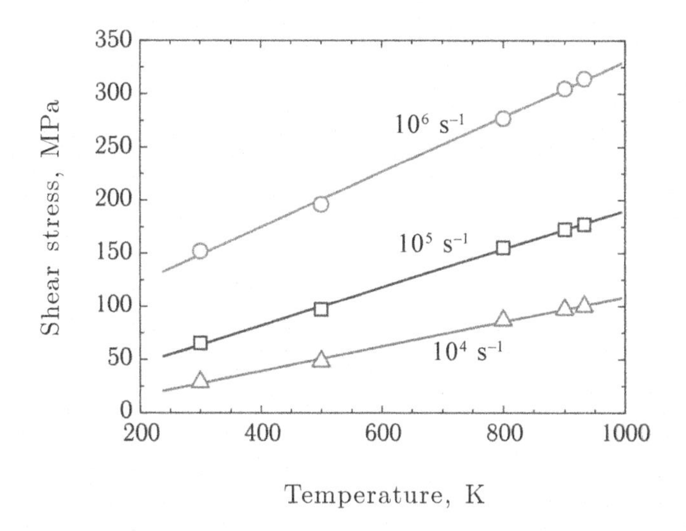

Fig. 5.20. Temperature dependences of the initial flow stress in pure aluminium at three fixed plastic strain rates. The linear character of the dependences is consistent with the phonon mechanism of dislocation dragging.

where $\alpha = 0.88 - 1.148\ (T/T_m) + 0.598\ (T/T_m)^2$ is the exponent in the empirical relation (5.4) for the precursor decay, varies from 0.33 to 0.6 ($T_m = 934$ K is the melting point of aluminium), and the coefficient A increases linearly with increasing temperature: $A = 0.01555T - 4.07$ with the dimensions of the magnitude of the stress MPa and the strain rate s^{-1}. In other words, the initial rate of plastic deformation behind the front of the elastic precursor increases with stress approximately in proportion to the third degree of the latter. In Fig. 5.20, the flow stresses in the elastic precursor from Fig. 5.18 are presented in the form of their temperature dependences with three fixed values of the plastic strain rate. The linear nature of these dependences is consistent with the mechanism of dragging of dislocations by phonon viscosity. On the other hand, a strong dependence of the initial rate of plastic deformation on stress contradicts the mechanism of phonon viscosity at a density of mobile dislocations corresponding to the initial state material.

The rate of plastic deformation is related to the density of mobile dislocations and their average velocity via the well-known Orowan relation (5.1). The values of the density of mobile dislocations of 10^7–10^9 cm^{-2}, which are necessary for ensuring the observed initial rates of plastic deformation, exceed the usual total dislocation density in the annealed pure metal of 10^6–10^7 cm^{-2}. Therefore, to

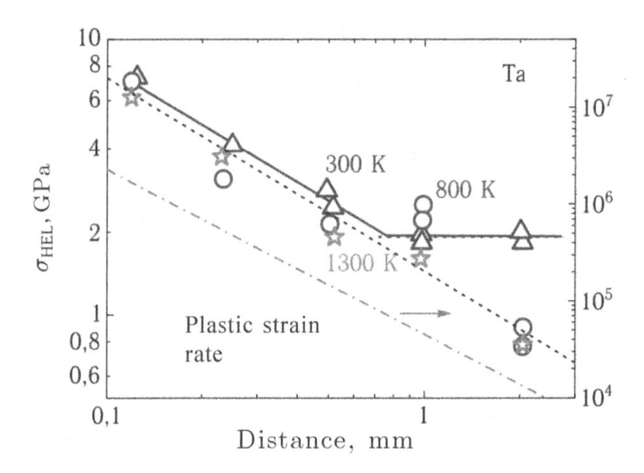

Fig. 5.21. Decay of the elastic precursor in tantalum at normal and elevated temperatures.

describe high-rate deformation in dislocation terms, it is necessary to assume their intensive nucleation or multiplication under the action of applied stresses. Similar results were obtained for silver and copper, as well as metals having the HCP structure. Molecular dynamic modelling of the movement of dislocations in FCC metals confirms a significant increase in the stress required to maintain their velocity with temperature, which is explained by an increase in phonon viscosity; mechanisms for rapid multiplication are not entirely clear.

On the other hand, BCC metals (iron [54], tantalum [51], vanadium [51]) exhibit somewhat different behaviour under these conditions. For an example Fig. 5.21 shows the results of measurements of the decay of the elastic precursor in high purity tantalum. Despite the considerable scatter of experimental data, which, apparently, is generally characteristic of metals with the BCC structure, measurements at room temperature clearly demonstrate a transition from the high-rate branch of the general dependence of the strain rate on the stress to the low-rate branch as the precursor decays. Of course, this transition also takes place for FCC metals, but it occurs at much lower stresses in the elastic precursor. Unlike FCC metals, in this case the anomalous increase in the Hugoniot elastic limit with increasing temperature is not recorded. The observed decrease in the Hugoniot elastic limit as a result of heating is particularly significant for samples with a thickness of more than 1 mm; for thin samples the effect of temperature is small. A weak temperature dependence was

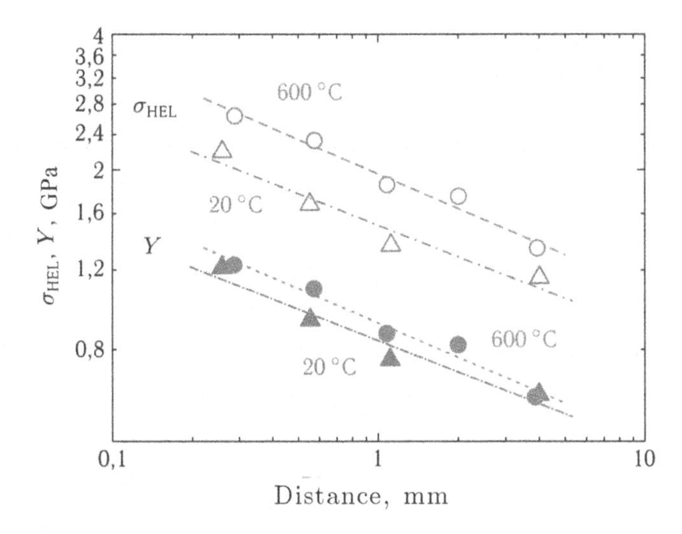

Fig. 5.22. Decay of the elastic precursor in VT1-0 titanium at normal and elevated temperatures. σ_{HEL} is the Hugoniot elastic limit equal to the stress in the elastic precursor, and Y is the plastic flow stress.

also obtained in molecular dynamics simulation [75] of the motion of dislocations at stresses above the Peierls stress; at lower stresses, the motion of dislocations has a pronounced thermal-fluctuation character. Considering the resistance of motion of dislocations as the sum of the contributions of the Peierls stress and phonon viscosity, the temperature dependence can be weakened as the result of opposite temperature effects in these components.

Metals with a hexagonal close-packed (HCP) structure are usually less ductile and, in their pure form, tend to form a coarse-grained structure, which makes it difficult to carry out similar studies with them. To date, experimental data have been obtained only for alloys (or metals with impurities) and single crystals.

Figure 5.22 shows the results of measurements [46] of the decay of the elastic precursor in the annealed commercial titanium VT1-0. The graph shows that the dynamic limit of elasticity increased slightly at elevated temperatures. However, this increase is insignificant, and the values of the flow stress, calculated from these data, taking into account the temperature dependences of the elastic moduli, are already practically insensitive to temperature. Figure 5.23 compares the dependences of the strain rate on stress in an elastic precursor and in a plastic shock wave. As for other metals, the strain rate in a plastic shock wave at the same shear stress is

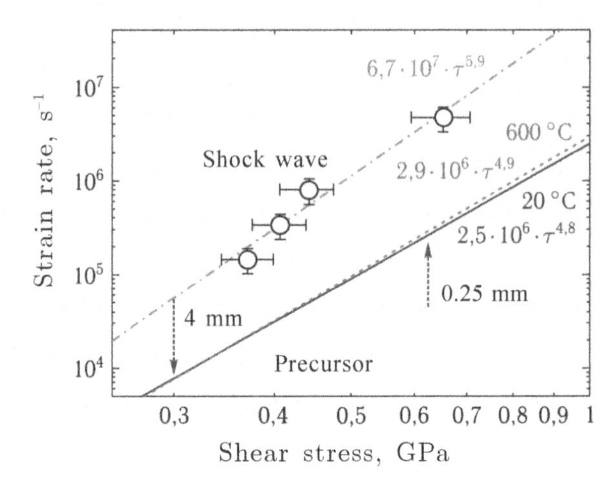

Fig. 5.23. Dependence of the plastic strain rate on the magnitude of shear stress in the elastic precursor for titanium VT1-0 at normal temperature (solid line) and at 600°C (dashed line). The vertical lines show the boundaries of the studied parameter range. Round markers connected by a dash-dotted line represent the strain rate in a plastic shock wave.

approximately an order of magnitude greater than the initial rate of plastic strain in an elastic precursor. In the state after rolling, the precursor decay is insignificant, the value of the Hugoniot elastic limit in the entire range of the distances travelled by the wave remains almost unchanged and corresponds to the HEL value in the annealed material at large distances, and the compression rate in the plastic shock wave is about the same as in the annealed material.

5.7. Behaviour of single crystals under shock-wave loading

Shock-wave studies of elastoplastic strains and the fracture of single crystals occupy a special place, since they make it possible to separate the contributions of various mechanisms for the nucleation and development of plastic strains and fracture. For obvious reasons, experiments with single crystals of metals are few. Probably, the largest amount of them was carried out with metals having a hexagonal close-packed structure: with beryllium, zinc, and magnesium. Single crystals of BCC metals – molybdenum and tantalum – and with a FCC structure – copper, aluminium – were also investigated, but the most interesting results were obtained in experiments with HCP crystals. Below are some results of experiments [55] with magnesium single crystals.

The mechanism of magnesium deformation includes three well-known slip systems: primary base, prismatic and pyramidal. If the base slip is suppressed, then twinning makes a significant contribution to plastic deformation. For shock-wave measurements, it is important that by changing the direction of wave propagation relative to the axes of the crystal, one can study the plastic flow in each system separately. The primary base slip should be observed under shock compression in the direction inclined to the *c* axis of the crystal. Secondary prismatic slip is activated when the wave propagates in the direction perpendicular to the *c* axis of symmetry of the crystal, when the shear stresses are absent in the primary basic slip system. Finally, the third, pyramidal, system of sliding and twinning should be activated when the shock wave propagates along the *c* axis of the crystal.

Figure 5.24 compares the wave profiles measured under shock-wave loading in the direction of the *c* and *a* axes of the crystal and at an angle of 45° relative to the *c* axis. The measurement results clearly demonstrate the dependence of the Hugoniot elastic limit and spall strength on the direction of application of the shock load. As expected, the slip along the (0001) basal plane occurs at the lowest value of shear stress and gives the smallest value of the Hugoniot elastic limit under shock compression.

Figures 5.25 and 5.26 demonstrate the effect of temperature on the value of the Hugoniot elastic limit of single-crystal magnesium

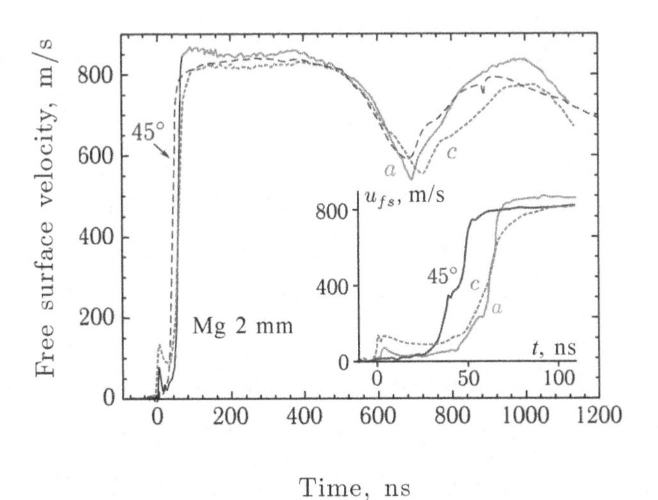

Fig. 5.24. The free surface velocity histories of single crystal magnesium samples with a thickness of 2 mm for three different orientations.

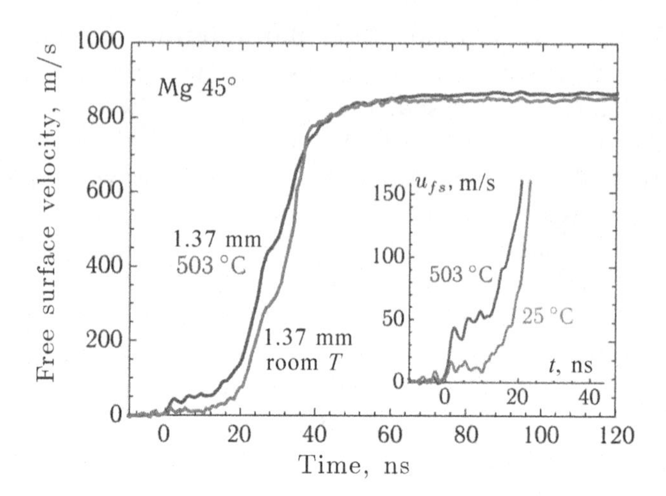

Fig. 5.25. The free surface velocity histories of samples with an orientation of 45° relative to the (0001) plane at normal and elevated temperatures.

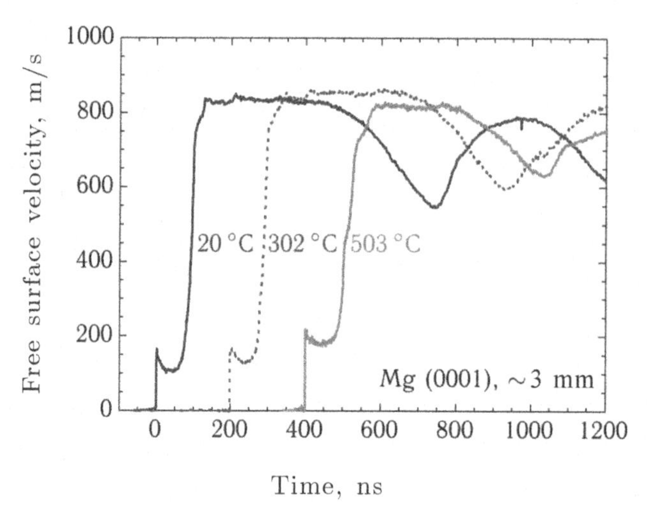

Fig. 5.26. The free surface velocity histories of samples with a thickness of 3 mm with orientation (0001) at normal and elevated temperatures.

samples of two orientations. It was expected that the phonon viscosity forces can be comparable with the flow stress in the basal plane, which should cause an increase in the Hugoniot elastic limit with an increase in temperature under shock compression in the inclined direction. The experiments confirm this assumption. The increase in the Hugoniot elastic limit with increasing temperature under shock compression in the direction of the *c* axis of the crystal, when plastic deformation occurs with the participation of twinning and the flow stress is large, was unexpected. The calculation of stresses in the planes of plastic shear, taking into account the temperature dependences of the elastic moduli, shows that the effect is much less but in any case, the high-rate plastic flow stress in all directions in the crystal does not decrease with heating.

An interesting feature of the presented wave profiles is the formation of two plastic waves propagating over a relatively large distance. For load orientations along and perpendicular to the symmetry axis of the crystal, an additional step on the plastic shock wave profile appears as a result of the reflections of the elastic precursor between the free surface and the plastic shock wave. However, this explanation cannot be true for samples with an inclined orientation, where the Hugoniot elastic limit is too small. It should be noted that the splitting of a plastic shock wave propagating at an angle of 45° to the crystal axis, which was also observed in experiments with beryllium single crystals, contradicts the theory of the propagation of plane waves in anisotropic elastoplastic media [56].

The formation of an additional step on the wave profile in a sample with a 45° orientation can be explained by the activation of higher order slip systems. Under uniaxial compression, sliding in inclined basal planes (0001) can provide relaxation of longitudinal and one of the transverse components of the deviator stresses, but cannot lead to relaxation of the third component with the normal in the z direction. For this reason, the stress difference $\sigma_x - \sigma_z$ continues to increase with compression, despite the fact that slip occurs in the (0001) plane, and as a result, the secondary slip system must be activated when the stress difference $\sigma_x - \sigma_z$ becomes sufficiently large. If two channels of stress relaxation are activated one after the other, it is natural to assume that the first plastic shock wave should be similar to an elastic precursor in the sense of a certain propagation velocity and a weak dependence of its parameters on the final shock compression pressure. In order to clarify the wave dynamics under shock compression in the inclined direction, additional experiments with varying shock compression pressures were carried out. The results are shown in Fig. 5.27. From the data presented it can be seen that the parameters behind the first plastic wave are the higher the higher the final shock compression pressure. A similar feature of wave dynamics was observed at cracking of shock-compressed glass in failure waves. The failure wave velocity is determined by the growth rate of cracks, very weakly depends on pressure and is

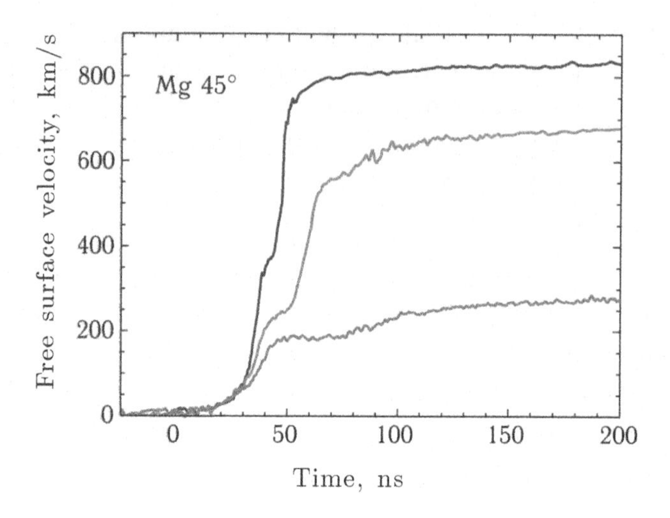

Fig. 5.27. The free surface velocity histories of samples with a thickness of 2 mm with an orientation of 45° at various pressures of shock compression.

not directly related to the compressibility of material, which means that the stress state ahead of the failure wave is determined by the parameters of the state behind it. By analogy with the failure wave, it can be assumed that an additional 'step' on the wave profiles of samples with an inclined orientation is associated with the propagation of the twinning front.

The wave configuration, exotic for metallic materials, without a leading elastic precursor was observed in experiments with zinc single crystals, the results of which are shown in Fig. 5.28. At a particle velocity of up to ~1 km/s (compression stress 14 GPa), the compression wave is actually a jump in parameters with a rise time of no more than 1–1.5 ns, followed by a section of constant parameters. With increasing pressure of shock compression behind the shock front, a section of relative smooth increase of the parameters appears, the duration of which decreases as the intensity of the shock wave increases. It is natural to assume that the appearance of this dissipative section means the beginning of plastic deformation in the shock wave. The reduction in the duration of the stress relaxation domain with increasing pressure of shock compression obviously has the same nature as the usually observed decrease in the width of plastic shock waves.

Thus, when the compression stress behind the shock front ≈14–15 GPa, there is a qualitative change in the response of the material to the load, which indicates a transition from purely elastic

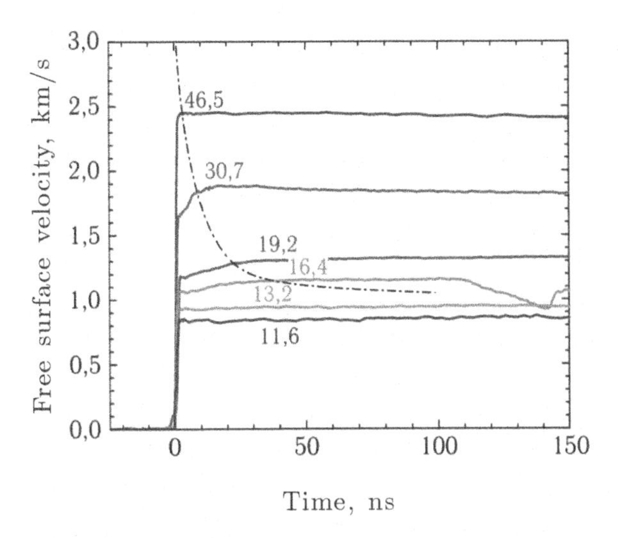

Fig. 5.28. The free surface velocity histories of zinc samples with a thickness of 0.25–0.5 mm with ⟨0001⟩ orientation at different pressures of shock compression. Shock waves with a pressure of more than 30 GPa were generated by the detonation of the explosive charge in contact with the screen; in the remaining cases, the plates collided. The maximum values of compressive stresses behind the shock wave front are indicated.

to elastoplastic deformation in this range of parameters without loss of stability of the shock wave. In steady compression waves, all intermediate and final states should correspond to the Rayleigh (Michelson) line

$$\sigma_x = \rho_0 U_S^2 \frac{V_0 - V}{V_0},$$

where V_0, V are the initial and current values of the specific volume of the substance. The positions of the Rayleigh lines with respect to the Hugoniot of polycrystalline zinc and the adiabat of elastic compression in the ⟨0001⟩ direction are shown in Fig. 5.29. In the absence of better data, the Hugoniot of polycrystalline zinc, obtained for a wide range of pressures, is considered here as the adiabat of all-round compression. On the Rayleigh lines, the final states and states corresponding to the onset of stress relaxation are marked. At an elastic compression stress of 14–15 GPa, corresponding to the onset of plastic deformation, the deviation of the Hugoniot of elastic longitudinal compression from the Hugoniot of zinc is ~1.5 GPa, which almost coincides with the value of the deviator

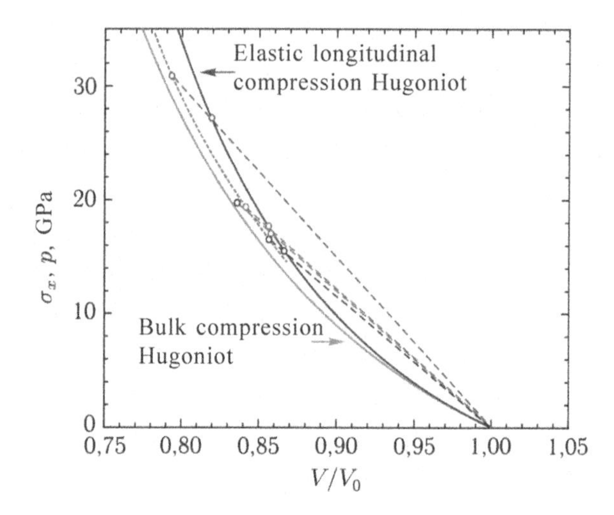

Fig. 5.29. Changes in the state of zinc in shock waves of varying intensity. The dots mark the final states and the states corresponding to the onset of stress relaxation.

stresses in elastic precursor waves propagating in the transverse direction. Deviations of the final states of shock compression from the Hugoniot of polycrystalline zinc are in the range of 1.2–2 GPa.

5.8. Dislocation multiplication effects

The strong dependence of the initial plastic strain rate behind the front of the elastic precursor on the stress contradicts the general dependence of the velocity of plastic flow on the stress shown in Fig. 5.1 expected for the high-rate branch. The expectations were based on the Orowan relation (5.1), in which the dislocation velocity, controlled in this mode by phonon viscosity, is a linear function of stress, and the density of mobile dislocations corresponds to the initial state of the annealed material and is constant. Strong nonlinear dependences of the initial plastic strain rate behind the front of an elastic precursor on stress and unexpectedly large values of this rate itself can be interpreted as evidence that the process is controlled more by the nucleation and multiplication of dislocations than by their speed. Probably, the nucleation and multiplication of dislocations can largely occur directly in the frontal part of the elastic precursor, in the process of compression, which is usually considered to be purely elastic.

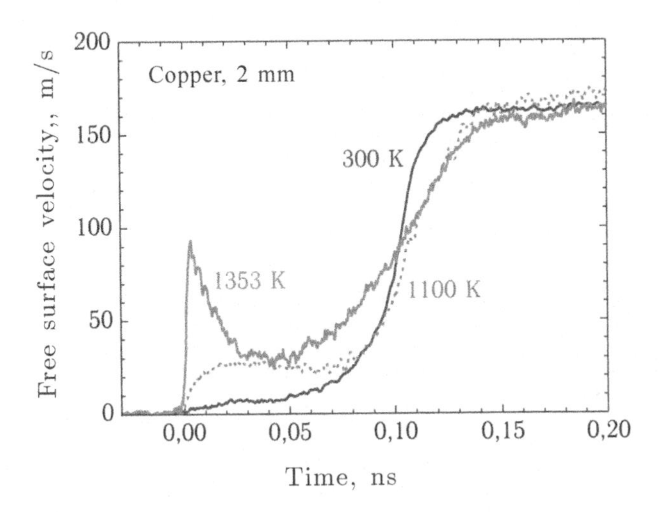

Fig. 5.30. The change in the Hugoniot elastic limit and the shape of the elastic precursor of the shock compression wave in pure copper with an increase in the test temperature [62].

Figure 5.30 shows the initial sections of the free surface velocity histories of 2 mm pure copper samples subjected to shock compression at different temperatures [57].

The relatively long rise time of the parameters at room temperature in the elastic part of the wave profile attracts attention. Usually, when analyzing the decay of the precursor, the rise time is neglected, considering this part of the wave as a shock discontinuity, but it is quite probable that with a long rise time the multiplication of dislocations takes place already during compression in the dispersed front of the precursor. Stress relaxation slows down with an increase in temperature and a corresponding increase in the contribution of phonon viscosity to the resistance to plastic deformation and this is manifested in an increase in the amplitude of the precursor and the rise time of a plastic shock wave. Due to the effects of nonlinearity with increasing compression stress in the elastic precursor, the time of growth of the parameters in it decreases: a sharp decrease in the width of the elastic wave near the melting temperature is also accompanied by a change in its shape. The stress peak in the frontal part usually appears due to accelerated stress relaxation, for example, as a result of intense multiplication of dislocations. From the comparison of the wave profiles in Fig. 5.30 it can be assumed that a reduction in the rise time shifts the intensive multiplication to the area immediately behind the elastic jump. Unfortunately, there is

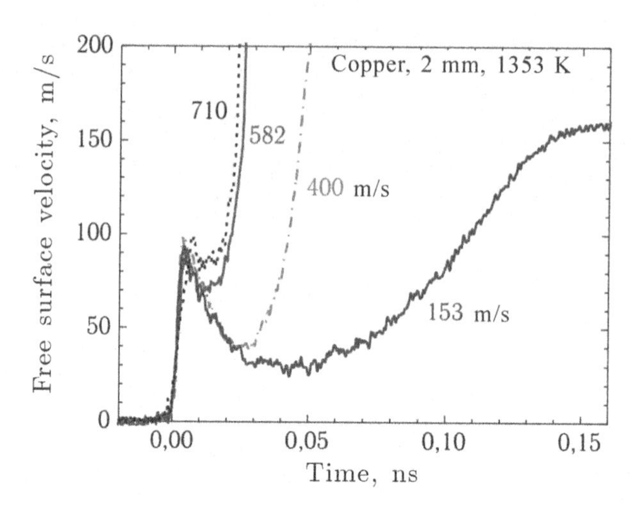

Fig. 5.31. The free surface velocity histories of 2 mm copper samples subjected to impact by a copper plate at different speeds at a temperature of 1353 K. The numbers at the histories indicate the impact velocity in m/s.

currently no complete theory of waves in relaxing media that would establish a quantitative relationship between the law of relaxation and wave evolution.

Figure 5.31 illustrates an important feature of the evolution of an elastic precursor with a stress peak in the frontal part. It can be seen that the parameters at the minimum between the elastic and plastic waves increase with an increase in the impact velocity, while the recorded value of HEL at the peak of the precursor remains almost unchanged. The section of the elastic precursor from its front almost to the minimum point is well reproduced regardless of the impact velocity. In other words, mechanical perturbations from a plastic shock wave cannot pass through a minimum point and have any effect on the frontal part of an elastic precursor.

According to the results of measurements of the evolution of elastic precursors in copper at a temperature of 1353 K, it was possible to estimate the rate of plastic deformation in the precursor at its various stages and then, using the Orowan ratio and calculated data on the dislocation drag coefficient, estimate the change in the density of mobile dislocations. The results are shown in Figs. 5.32 and 5.33. These estimates showed that the dislocation density increases by almost an order of magnitude in the first 13–15 nanoseconds after compression at the front of the elastic precursor and then remains almost constant, while the state of the material

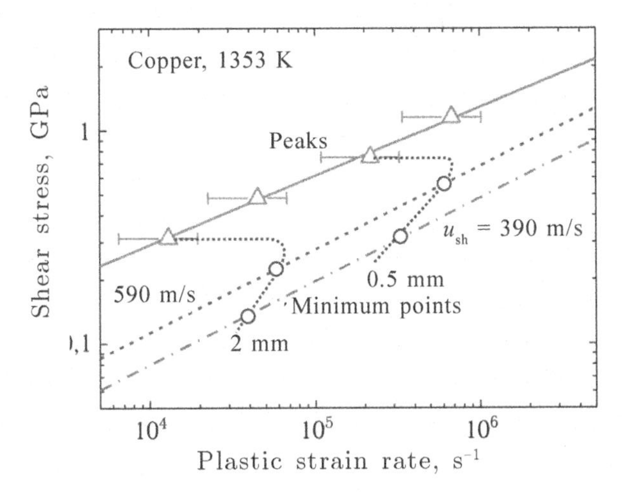

Fig. 5.32. The ratio between the plastic strain rate and the stress in a decaying elastic precursor in copper at a temperature of 1353 K [62]. Dot lines connect points for wave propagation distances of 2 mm and 0.5 mm.

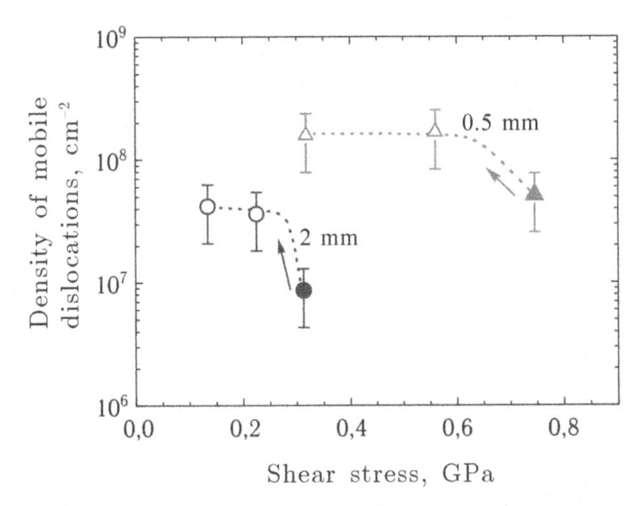

Fig. 5.33. The density of mobile dislocations as a function of shear stress according to Fig. 5.32. Solid markers show the parameters at the peak of the precursor. The arrows indicate the direction of change of parameters beyond the maximum stress at the peak.

approaches the minimum point between elastic and plastic waves. The plastic strain itself in the first 13–15 nanoseconds reaches about 0.1%. Termination of the multiplication of the dislocations during a stress drop can be interpreted as evidence that multiplication is controlled not only by the value of deformation, but also by the magnitude of the effective stress. The subsequent compression in the

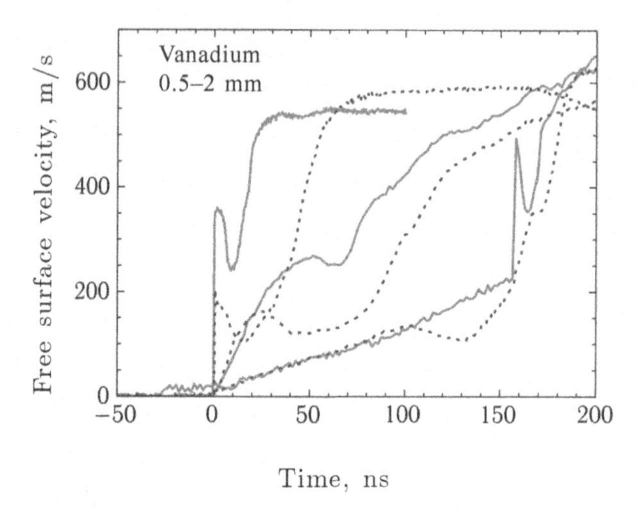

Fig. 5.34. The free surface velocity histories of vanadium samples with a thickness of 0.5 mm (solid lines) and 2 mm (dashed lines), measured (from left to right) with shock, shockless and mixed dynamic compression.

plastic shock wave is accompanied by an increase in the shear stress and, as already mentioned, by further multiplication of dislocations.

Since the experimental data obtained suggest that significant multiplication of dislocations may occur already in the process of compression in the frontal part of the elastic precursor, it was natural to investigate the relationship between the rise time of the parameters in the elastic precursor and the transient value of the Hugoniot elastic limit, as well as the general evolution of the elastoplastic shock compression wave. For this purpose, experiments with vanadium were carried out [58], in which a compression wave was introduced into the samples through intermediate plates of different types of glass. The compressibility of silicate glasses behaves in an abnormal way, as a result of which the compression waves in them increase in width as they propagate. Figure 5.34 compares the wave profiles measured for samples with a thickness of 0.5 mm and 2 mm with three types of dynamic compression – shock, shockless and mixed. The final compressive stress in all three series was approximately the same. The parameters of the mixed compression are such that the transition from the ramped to the shock-wave part occurs at a stress less than the Hugoniot elastic limit at large distances. However, the increase in compression time in this series from 1–2 ns to about 150 ns did not lead to a significant stress drop at the peak of the precursor. The recorded parameters at the peak for a sample with

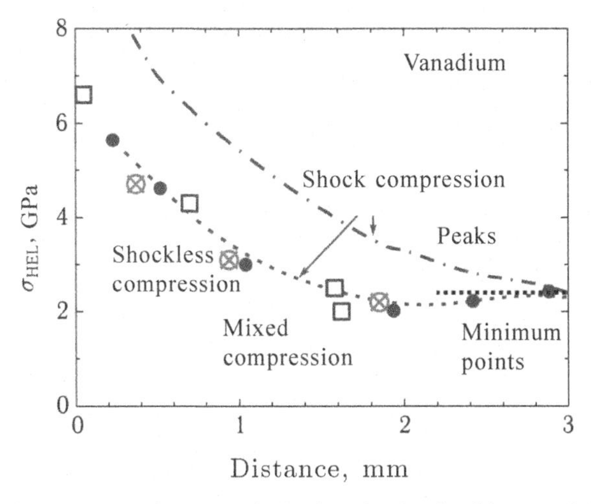

Fig. 5.35. Decay of an elastic precursor during shock, shockless, and mixed dynamic compression of vanadium. Markers show the parameters at the minima between the elastic and plastic waves, the dashed–dotted line describes the decay of the precursor peak under shock compression, the dotted line on the right shows the estimate of the HEL value after the end of the precursor decay.

a thickness of 0.5 mm are even higher in this case than in shock compression. The latter is not explained by the stress relaxation feature, but by the wave dynamics of reflection of the compression pulse from the free surface. On the other hand, in the experiments without shock compression, the stress at the peak of the precursor actually decreased due to an increase in the rise time.

More interesting is the result of this series of experiments, shown in Fig. 5.35. It turned out that for all three variants of dynamic compression, the parameters at the minimum between the elastic and plastic waves evolve as they propagate almost equally and can be approximated by a single functional dependence. This means that, despite the different deformation histories, the material approaches the minimum point with practically the same density of mobile dislocations.

The graph in Fig. 5.35 also demonstrates the nonmonotonic evolution of an elastoplastic shock compression wave in annealed vanadium, which has not been observed previously. The decay of the elastic precursor is caused by the relaxation of stresses behind its front, but no relaxation process can lead to an increase in the stress in the wave as it propagates. A possible mechanism of stress growth in the minimum between elastic and plastic waves follows

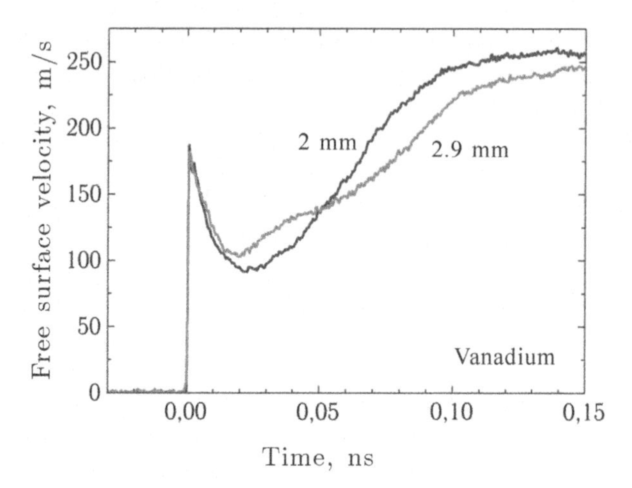

Fig. 5.36. The free surface velocity histories of the annealed samples, measured in the region of the nonmonotonic evolution of the elastoplastic shock compression wave. The numbers in the histories show the sample thickness.

from a comparison of the wave profiles in Fig. 5.36, which clearly demonstrates the plastic shock wave emission of a new elastic precursor after it has travelled a distance greater than 2 mm. This means that the stress ahead of the plastic shock wave in the process of decay of the precursor fell below the elastic limit of the material. The newly emitted elastic wave raises the stress before the plastic shock wave to the current value of the Hugoniot elastic limit. Anyway, the detected non-monotonicity is a consequence of a certain ratio of the contributions from the multiplication of dislocations and the relaxation of stress caused by them to the waveguide dynamics. On the other hand, it is known that twinning makes a large contribution to the mechanism of plastic deformation of vanadium. From this point of view, the observed feature can be related to the fact that the nucleation of twins requires a significantly greater stress in the bulk of the material than for their growth.

6

Shock wave compression of hard brittle materials

Very hard brittle materials are widely used in science and technology, including under conditions associated with the action of large static and dynamic compressions. Special attention has been paid in recent decades in connection with a number of practical applications of the behaviour of hard brittle materials (rocks, ceramics, glasses) under shock loading.

Brittle materials typically have a tensile strength that is much less than the elastic limit and, unlike ductile materials, can also be fractured under compression. The basic patterns and mechanisms of fracture of brittle materials under compression are the subject of many investigations, but to date have not been fully studied. Experiments with shock waves in this sense are attractive by specific loading conditions, characterized by strictly one-dimensional deformation and a practically unlimited range of attainable stresses.

By definition, fracture means a discontinuity of the material. Cracks, like other discontinuities, can be formed in a non-porous medium only under the action of tensile stresses. It is known, however, that even with general compression, local stresses near inhomogeneities can become tensile. One of the possible mechanisms for generating tensile stresses under compression is schematically shown in Fig. 6.1. It is assumed that the material has inclined cracks or other similar defects of limited size, along which shear strains are facilitated. In non-hydrostatic compression, the sliding of the material along the surfaces of the inclined crack forms areas of excessive compression and tension at its ends (Fig. 6.1 *a*). Under certain

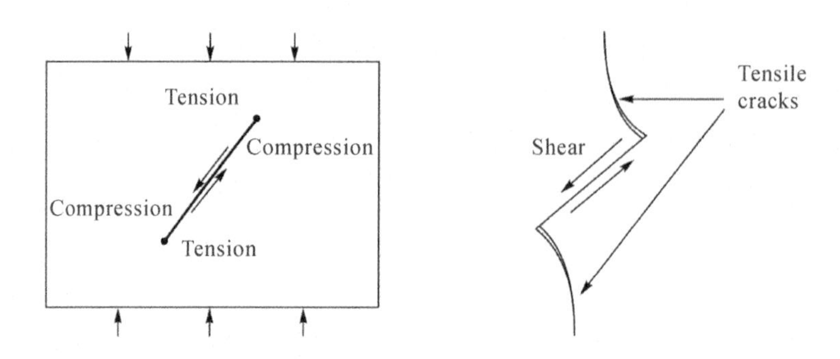

Fig. 6.1. Scheme of formation of wing-shaped cracks during non-hydrostatic compression.

conditions, the stress is sufficient to initiate fracture. According to the Griffith criterion [59], fracture in compression is triggered when the highest local tensile stress reaches the threshold value. For a biaxial stress state, the Griffith criterion is

$$(\sigma_1 - \sigma_2)^2 + 8K(\sigma_1 + \sigma_2) = 0, \tag{6.1}$$

where σ_1, σ_2 are the main stresses, K is a material constant, which is assumed to be equal to the standard tensile strength.

Model experiments with glass and polymer plates showed that a local shift along the surface of an inclined crack leads to the formation of a configuration of three cracks, called a wing-shaped crack (Fig. 6.1 *b*). At the edges of the local shear region, tensile cracks are initiated that grow along a curved surface outside the shear plane, deviating in the direction of compression. As the growth direction approaches the compression direction, the process slows down to a full stop, however, the crack growth can resume when the compressive stress increases. The beginning of the growth of cracks under compression does not immediately lead to a loss of the mechanical stability of the body, in contrast to how this occurs under tension. Lateral pressure prevents the opening of axial cracks and thereby increases the threshold of fracture until the onset of plastic flow.

It must be said that there are no absolutely brittle materials. All materials that are brittle under the normal conditions, including even diamond, show noticeable plasticity under the action of high pressure. However, unlike ductile metals, high-hardness oxides and intermetallic compounds with covalent–ionic interatomic bonds and low symmetry of crystals are characterized by a high energy of dislocation formation and a small number of planes, where dislocation slip is possible. For this reason, plastic deformation is very difficult, as a result of which a high concentration of stresses on microcracks and other inhomogeneities is possible. Blocking the slip inside the body at the grain boundary or at the intersection with another acting shear system can lead to the appearance of a crack.

When using the results of shock-wave measurements, one should take into account the fact that one-dimensional compression in a shock wave is associated with an increase not only of longitudinal, but also of transverse stresses. Since plastic deformation and fracture by compression have different physical nature and, accordingly, differently depend on the existing stresses, to build the constitutive relations and models describing the resistance to inelastic deformation of brittle materials, it is important to know whether brittle fracture under shock compression of the test material actually took place.

Studies of the behaviour of brittle materials under shock-wave loading include measurements of shock compressibility, recording and analysis of waveforms of compressive stress or the velocity of a substance, measurements of the stress state in a shock-compressed material, and spall strength after loading with pulses of varying intensity. The most controversial was the question of the nature of inelastic deformation in a shock wave, namely: does brittle cracking or plastic flow take place during compression of a brittle material in a plane shock wave. The fact is that with uniaxial shock compression both longitudinal and transverse stress components increase. In the elastic region, the change in the longitudinal σ_x and transverse σ_y stresses occurs in a consistent manner:

$$\sigma_x = \sigma_y (1 - v) / v,$$

where v is the Poisson's ratio. The fracture threshold quickly increases with increasing transverse compressive stress, and at a certain value of σ_y a so-called brittle–ductile transition occurs: shear stresses become sufficient to activate plastic deformation mechanisms, and crack opening is suppressed by transverse stresses. Resistance to inelastic deformation when destroyed on the one hand and plastic flow on the other have a different physical nature and are described in various ways. For this reason, in order to calibrate rheological models, it is extremely important to know the nature of the observed inelastic deformation.

6.1. Shock compression and fracture of glass Failure waves

One of the traditional model objects for the study of fracture patterns is glass. Under shock compression, its behaviour is distinguished by a number of specific features, the most obvious of which is the anomalous compressibility in the elastic region. An abnormal decrease in the speed of sound during compression leads to the fact that the elastic compression waves in glass expand as they propagate. As seen in Fig. 6.2, the area of abnormal compressibility in different glasses occupies different pressure ranges (compressive stresses). At the same time, for a number of glasses – the optical crown K8, soda-lime glass, etc., the transition from elastic to inelastic deformation becomes blurred and difficult to determine. Since elastic compression is reversible, an anomaly of elastic compressibility must lead to the formation of a shock rarefaction wave. Figure 6.3 shows that this is indeed the case.

Exceeding the elastic limit is accompanied by a sharp increase in the speed of sound and the disappearance of the conditions for the formation of a shock rarefaction wave. This can be seen from the results of measuring the evolution of the shock compression pulse in K8 glass, shown in Fig. 6.4. The measurement results clearly demonstrate the irreversible densification of glass, which is manifested in a significant residual deformation after the passage of a compression pulse. The magnitude of the irreversible densification of

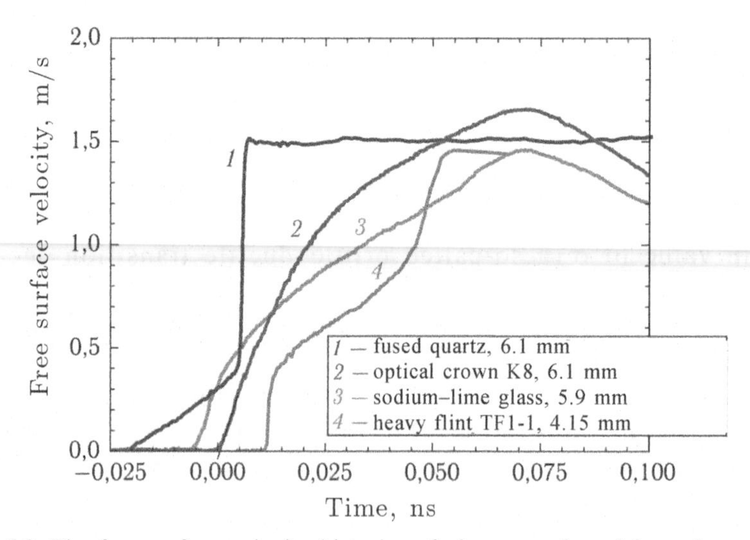

Fig. 6.2. The free surface velocity histories of plane samples of four glass grades.

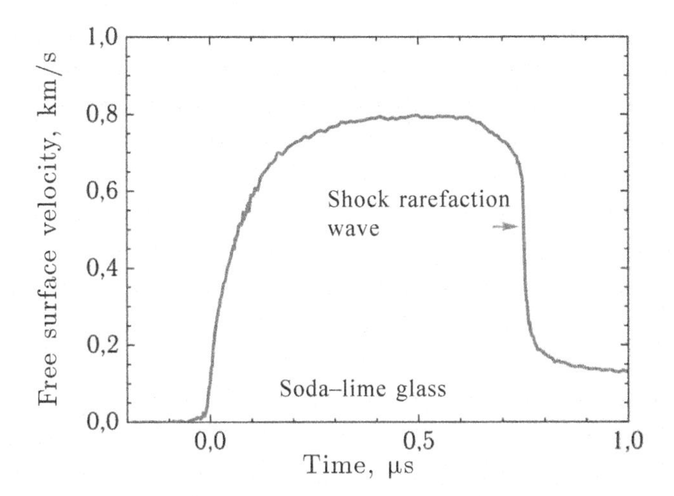

Fig. 6.3. The formation of a shock rarefaction wave during unloading in soda–lime glass. Experiment with a water window.

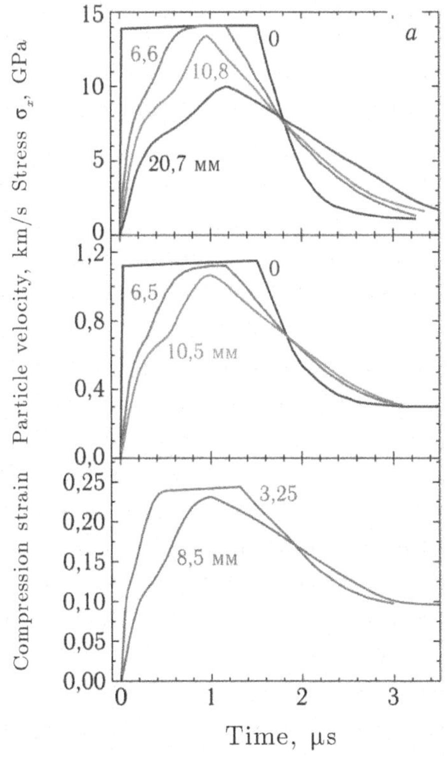

Fig. 6.4. The histories of the longitudinal stress $\sigma_x(t)$, the particle velocity (particle velocity) $up(t)$, and the strain $\varepsilon_x = 1 - V/V_0$ under shock compression of K8 glass. The distances from the impact surface at which the gauges were installed are indicated.

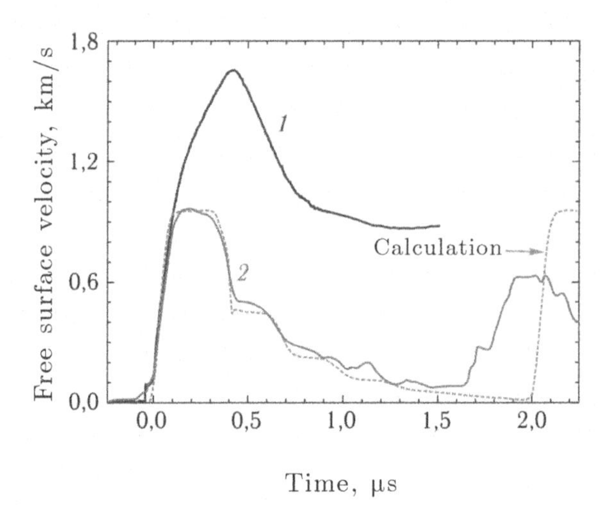

Fig. 6.5. The free surface velocity histories of the K8 optical glass at two impact speeds. In experiment 1, a paraffin layer was placed behind the aluminium flyer plate, which limited unloading after shock compression. The dotted line shows the calculated velocity history obtained as a result of computer simulation of experiment 2 without taking into account the damage.

K8 glass after shock compression to 14 GPa is 7.5–8%. Irreversible densification under the action of high pressure is characteristic of silicate glasses with a loose structure. In particular, the irreversible densification of fused quartz reaches 15% at shock compression pressures of 10–15 GPa. It is assumed that the irreversible densification of glass underlies the mechanism of its plasticity in compression.

Figure 6.5 shows the velocity profiles of the free surface of K8 glass samples, measured under shock compression conditions above and below the Hugoniot elastic limit. In both cases, the measured wave profiles repeat the shape of the compression pulse in the sample. The stepped nature of the decrease in velocity in a rarefaction wave at a lower impact velocity is explained by the difference in the dynamic impedances of glass and a steel impactor. The incomplete unloading in the experiment with greater intensity of the shock wave is explained by the fact that in this experiment a thick layer of paraffin adjoined the back surface of the impactor.

The release of the spall pulse to the surface on the presented wave profiles is not recorded. It follows that the dynamic tensile strength of the glass exceeds 6.8 GPa under shock compression below the elastic limit and remains extremely large when the latter is exceeded.

For comparison: the quasistatic tensile strength of glass is about 0.1 GPa. The reason for such a large discrepancy is that, under normal conditions, the fracture of glass is initiated on its surface, where there are always nucleation microcracks. At the same time, in experiments with shock waves, spall fracture can be initiated only in the bulk of the material without any involvement of surface defects. Maintaining high glass strength when the Hugoniot elastic limit is exceeded means that the material is not destroyed by compression, the elastic limit is associated with the onset of plastic deformation, and the plasticity of the material is maintained during the unloading process from the shock-compressed state and subsequent stretching.

In Fig. 6.5 the measured free surface velocity history of a glass plate is compared with the result of a computer simulation of a steel plate impact on a plate of elastic non-destructive glass. It can be seen that the reverberation time of the compression/tension pulse in the glass plate is shorter than that expected for the longitudinal elastic wave. The fact is that the reflection of the tension wave in glass did not occur on the impact surface, but on the boundary of the surface layer that is in contact with the impactor, no more than 1.5 mm thick, fracture by cracks propagating from the surface. Such cracking does not occur when shock compression is significantly higher than the elastic limit. Experiments with samples of glass of different thickness, loaded with long duration shock compression pulses, showed that the fractured layer expands with time. This process can be represented as the propagation of a failure wave.

The term 'failure wave' appeared in the 1960s, when a hypothesis was put forward about the possibility of the process of material fragmentation in a relatively thin layer spreading at the speed of sound. As it propagates, this fracture front continually nucleates many new cracks in the intact material. In the late 1980s, the formation of failure waves in glass under the conditions of one-dimensional compression by a plane shock wave was experimentally detected. Subsequently, the fact of the formation of failure waves was repeatedly confirmed and an extensive amount of empirical information on the kinematic laws of their propagation and the limits of initiation was collected. The failure wave is a network of cracks initiated by the applied stress on the glass surface, where there are always numerous nucleation microcracks and propagating into the bulk of the material. The speed of propagation of a failure wave is less than the longitudinal and shear sound velocities, is close to the limiting crack growth rate (~1.5 km/s for glass) and depends on the stress. Failure waves are generated at compressive stresses above a certain threshold, which can be identified as a failure threshold. When the effective stress decreases, the failure wave stops. In the failure wave, there is a consistent increase in the longitudinal compressive stress and the material density in accordance with the laws of conservation of mass and momentum, and shear stress relaxation occurs. The failure wave velocity slightly increases with increasing compression stress. After passing the failure wave, the material completely or almost completely loses its resistance to tension. Plastic deformation suppresses cracking of the material.

Although it is shown that the failure wave is indeed a wave, as understood in continuum mechanics, its kinematics differs from the kinematics of elastic–plastic waves. A shock wave in an elastoplastic body loses its stability due to a sharp increase in compressibility upon reaching the yield point. As a result, the shock wave splits into elastic and plastic compression waves. The stress behind the front of the elastic wave is determined by the magnitude of the yield strength of the material. Such a wave structure should be formed in a polycrystalline brittle material, where the fracture is initiated in each grain as soon as the stress reaches the fracture threshold. In both cases, the velocity of propagation of the second wave is determined by the bulk compressibility of the material. The speed of propagation of a failure wave is determined by the rate of growth of cracks, which is not related to bulk compressibility. On the other hand, the final compression stress behind the failure wave is determined by the conditions on the impact surface. As a result, since the velocity of propagation of the failure wave and the final state of the material behind it are fixed, the stress ahead of its front (that is, the stress in the leading elastic wave) is determined by these conditions and is not necessarily equal to the fracture threshold.

The surface of the glass is a source of cracks and plays an important role in the wave process of fracture. Probably the most obvious is the role of surfaces and the specific kinematics of failure waves occur when compression waves propagate through a stack of glass plates. When passing through each surface, the compression wave splits into a leading elastic wave and a low-speed failure wave that follows. As a result, the stress at the front of the leading elastic wave decreases in a stepwise manner on each surface in the stack. This should occur until the stress in the elastic wave decreases to the fracture threshold. Figure 6.6 compares the results of measurements [60] of wave profiles generated by impact in a thick glass plate and

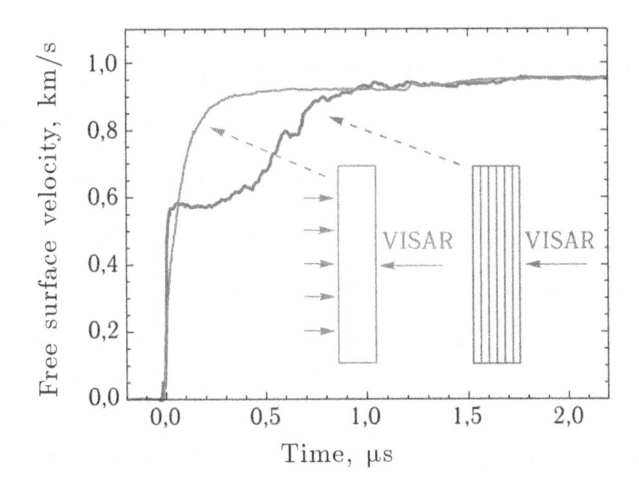

Fig. 6.6. Transformation of shock compression pulse in glass due to the formation of failure waves. The free surface velocity histores of a plate of soda–lime glass with a thickness of 5.9 mm and stacks of 8 glass plates with a thickness of ~1.21 mm each are presented under the same loading conditions.

in a stack of thin plates. The superposition of failure waves in a stack of glass plates forms a two-wave compression configuration. In this case, the time of compression in the second wave approximately corresponds to the time of propagation of counter cracks through the plate in the stack. The stress behind the front of the leading elastic wave in the stack is 4 GPa. Apparently, this stress is close to the threshold for the fracture of glass in these conditions. The final values of the velocity of the free surface are almost the same for a thick glass plate and a stack of thin plates. Thus, experiments with shock compression of a stack of plates are a simple and visual way to detect failure waves.

While it has been reliably established that the threshold for the formation of failure waves is below the elastic limit of glass, the position of the upper boundary of the range of existence of failure waves is not entirely clear. Figure 6.7 presents data [61] on the velocity of propagation of a failure wave as a function of compressive stress in two kinds of glass. It can be seen from the above data that the transition through the Hugoniot elastic limit is not accompanied by the appearance of any features on the dependence of the failure wave velocity on the magnitude of the compressive stress. The results of the experiments with two kinds of glass confirmed the results of experiments with pre-stressed glass samples [62] that

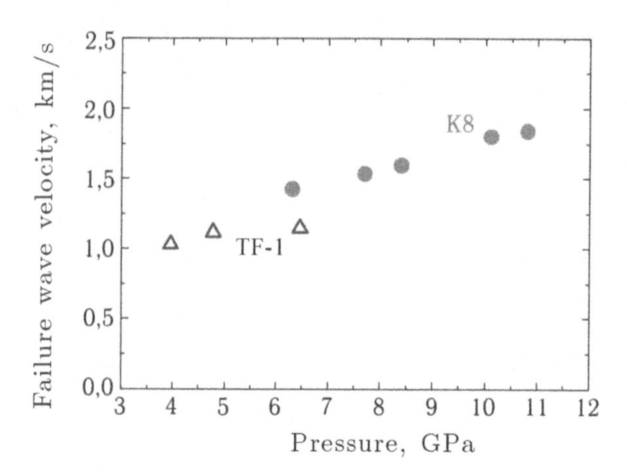

Fig. 6.7. The dependence of the velocity of failure waves in K8 and TF-1 glasses on compressive stress.

the upper limit of the stress range at which the formation of failure waves is possible significantly exceeds the elastic limit. Probably, in the mechanism of propagation of a failure wave, an important role is played by the processes of local densification in the vicinity of inclined cracks, and the upper threshold for the fracture of glass is determined by the beginning of its densification in the volume. The lower threshold for the results of experiments with stacks of glass plates was estimated as 5.3 GPa for K8 glass and 4 GPa for TF1.

6.2. Attempt to record failure waves during shock compression of ceramic materials

Observation of failure waves under uniaxial shock compression could be used to diagnose fracture under these conditions. Figure 6.8 presents the results of experiments with Al_2O_3 and B_4C ceramics. Unlike the glasses, in experiments with stacks of ceramic plates, the amplitude of the elastic precursor does not decrease due to the formation of failure waves. In fact, the particle velocity and, accordingly, the compressive stresses behind the front of the elastic precursor in experiments with stacks turned out to be even slightly higher than in experiments with monolithic samples. This increase is the result of the presence of thin gaps between the ceramic plates. The gaps interrupt the precursor decay and force it to resume from elevated stresses.

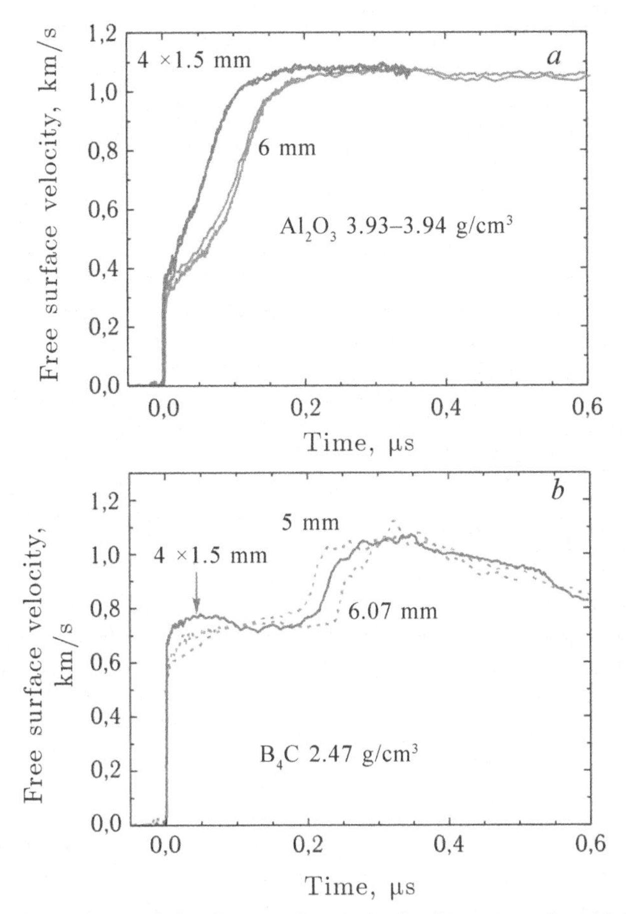

Fig. 6.8. Comparison of the free surface velocity histories of a thick plate and a stack of thin plates for ceramic alumina (*a*) and boron carbide (*b*). Loading by impact of an aluminium plate 2 mm thick with a speed of 1.9 km/s.

Thus, it is not entirely clear whether failure waves can form in other brittle materials other than glass. Probably, the initiation of compression failure waves can occur only in homogeneous materials, in which the concentration of nucleation microcracks in the surface layers is possible, while maintaining the uniformity and practical non-defectiveness of the material inside the body.

In any case, from the fact that failure waves in ceramics are not fixed, it does not at all follow that they are deformed under shock compression by the plastic mechanism.

6.3. Diagnosing the nature of the inelastic deformation of brittle materials under shock compression

Study [63] proposed a method for diagnosing the nature of inelastic deformation under shock compression of brittle materials by varying the magnitude of the transverse compressive stress and measuring its effect on the Hugoniot elastic limit. The idea of the method is illustrated in Fig. 6.9, where the relationship between the longitudinal σ_1 and transverse $\sigma_2 = \sigma_3$ stresses under uniaxial elastic compression $\sigma_1 = \sigma_2\,(1-\nu)\,/\nu$ (where ν is Poisson's ratio) is shown, as well as the boundaries of the elastic deformation region beyond which plastic flow or fracture under compression begins. The stresses are positive when compressed. The intersection of the wave beam $\sigma_1 = \sigma_2(1-\nu)/\nu$ with one of the lines describing the yield strength or fracture threshold determines the elastic limit for uniaxial compression.

It is known that in the brittle fracture domain the elastic limit strongly depends on pressure, while with the onset of plasticity, this dependence practically disappears. The fact is that the mechanisms of inelastic deformation during fracture and plastic flow have a different nature and are described in various ways. During plastic deformation, the elastic precursor amplitude must meet the yield criterion, for example, the von Mises or Tresco criteria, according

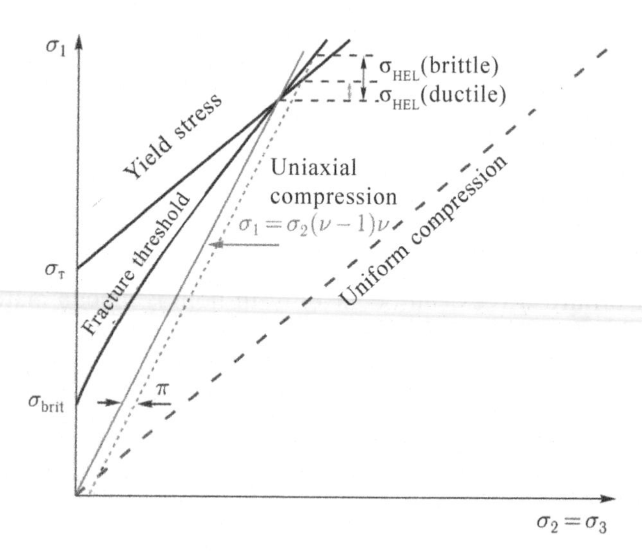

Fig. 6.9. Different effects of transverse stress on yield stress in uniaxial compression of ductile and brittle materials.

to which the stress on the Hugoniot elastic limit σ_{HEL} is related to the yield strength σ_T by the ratio $\sigma_{HEL} = \sigma_T\,(1-\nu)/(1-2\nu)$, where ν is Poisson's ratio.

The presence of a relatively small lateral pressure in ductile material π causes a small increase in the amplitude of the precursor: $\sigma^{duct}_{HEL} = (\sigma_T + \pi)\,(1-\nu)/\,/(1-2\nu)$. In the case of brittle behaviour, we can use the Griffiths fracture criterion, which gives $\sigma_{HEL} = \sigma_{brit}(1-\nu)/(1-2\nu)^2$. In this case, the imposition of lateral pressure leads to a much larger increase in the amplitude of the elastic precursor:

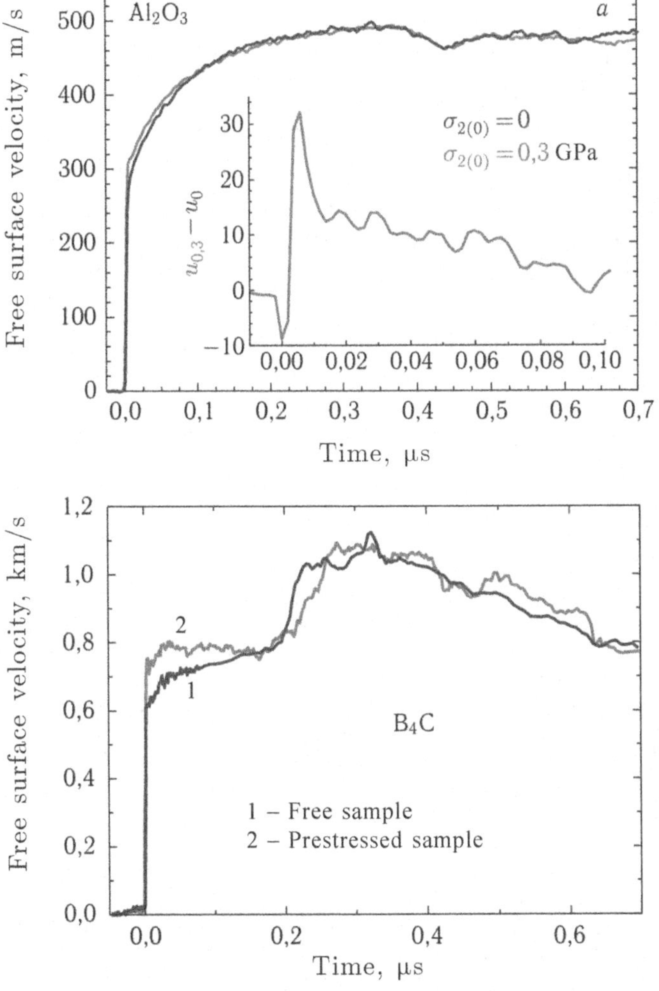

Fig. 6.10. Comparison of the free surface velocity histories of free (dashed curves) and prestressed (solid curves) ceramic alumina plates (graphs a) and boron carbide (b). Inset of the graph of a shows the change in the velocity difference with time.

$\sigma^{brit}_{HEL} = [\sigma_{brit} + (1-2v)(3-2v)\pi]\,(1-v)/(1-2v)^2$, which is about two and a half times the effect of lateral pressure in the plastic behaviour.

Figures 6.10 *a, b* present the measured [63] surface velocity profiles of free and prestressed aluminium oxide and boron carbide samples under shock-wave loading. Controlled lateral pressure $p \approx 0.3$ GPa in the samples was created by the method of hot fitting in steel rings, which initially had a smaller (by 0.1 mm) inner diameter than the samples. The results of the experiments shown in Fig. 6.10, clearly demonstrate the difference in the reactions of alumina and boron carbide on the lateral pressure. Measurements show that alumina behaves like a ductile material under one-dimensional compression in a shock wave, while in boron carbide there is a brittle fracture by compression.

Establishing the fact that uniaxial shock compression of boron carbide ceramics causes its fracture introduces certainty in the choice of the constitutive relation for describing the response of a material to a load. Figure 6.11 presents the results of numerical simulation of the wave process in a ceramic plate with a thickness of 8 mm, which was hit by an aluminium plate with a thickness of 2 mm and a speed of 1.9 km/s. In the calculations, the elastic deformation region is limited by the fracture threshold σ_f:

$$|\sigma_1-\sigma_2| \le \sigma_f,\ \sigma_f = \sigma_{f0} + f\sigma_2. \tag{6.2}$$

The behaviour of the fractured material is determined by the interparticle friction forces σ_c in accordance with the Coulomb – Mohr criterion:

$$|\sigma_1-\sigma_2| \le \sigma_c,\ \sigma_c = \sigma_{c0} + c\sigma_2, \tag{6.3}$$

$\sigma_{f0} = 5.8$ GPa, $f = 1.9$, $\sigma_{c0} = 1$ GPa and $c = 0.4$ are material constants. When the threshold σ_f is reached, the material begins to fracture. The degree of fracture is described by the parameter D: $0 < D < 1$. As the fracture proceeds, the threshold stress changes:

$$\sigma_D = (1-D)\sigma_f + D\sigma_c. \tag{6.4}$$

In the process of inelastic deformation, an accumulation of damage occurs in accordance with the ratio

$$\frac{dD}{d\varepsilon_p} = \varepsilon_p^f, \tag{6.5}$$

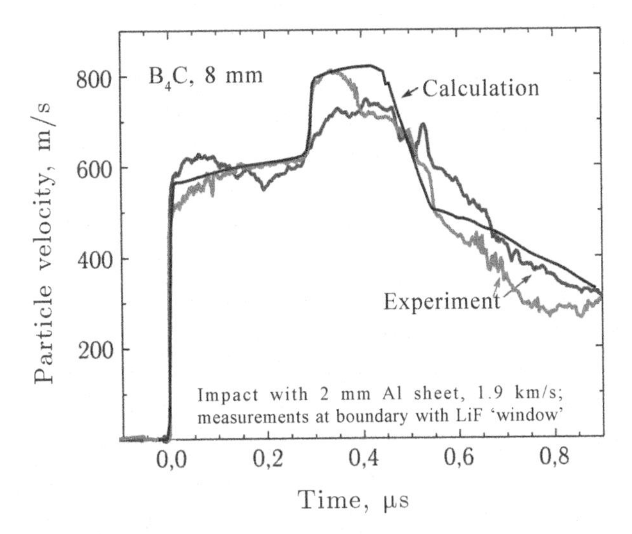

Fig. 6.11. The result of computer simulation of the shock-wave process in the sample of ceramics B_4C in comparison with the wave profiles measured at the boundary between the ceramic plate and the window of a LiF single crystal.

where ε_p^f is inelastic deformation at fracture at constant density of the material. In turn, the increase in the value of D leads to a decrease in the fracture threshold σ_D from the initial value σ_f to the value of σ_c determined by the forces of interparticle friction, which ensures the further development of the process.

When the stress difference is $|\sigma_x - \sigma_y| \leq \sigma_D$ response of the material to a change in load is elastic, but if the stress difference exceeds the threshold, then its relaxation occurs to σ_D. The relation for the description of stress relaxation is obtained from the known squared dependence of the strain rate $\dot{\varepsilon}_p$ on the shear stress τ in shock waves

$$\dot{\varepsilon}_p = \left(2\tau - \sigma_D\right)^2 / \eta'. \tag{6.6}$$

where $\eta' = 5 \cdot 10^{11}$ Pa^{2}s.

In Fig. 6.11 the calculation results are compared with the wave profiles measured in two independent experiments at the boundary between a ceramic sample and a single crystal of lithium fluoride, used as a 'window' for laser interferometric velocity measurements. Experimental wave profiles contain irregular oscillations, indicating a substantially heterogeneous inelastic deformation, but in general, the agreement between the measured and calculated profiles is quite satisfactory. Figure 6.12 shows the calculated trajectory of the state

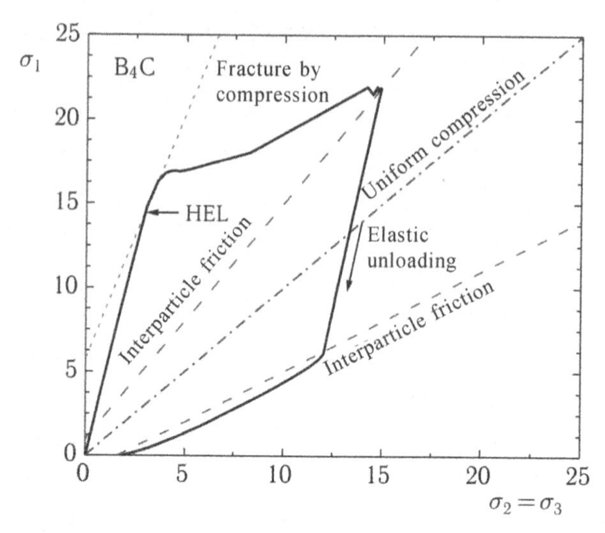

Fig. 6.12. The trajectory of the state change (solid line) under uniaxial compression and subsequent unloading of B_4C ceramics, obtained in the numerical experiment shown in Fig. 6.11. HEL – elastic limit with uniaxial shock compression.

change of the selected layer of the ceramic plate in the coordinates σ_1–σ_2.

6.4. Divergent shock compression

While the conditions of uniaxial elastic shock compression are characterized by a constant ratio between longitudinal and transverse stresses, in most practically important cases shock or explosive loading is associated with a variation of this ratio over a wide range. Figure 6.13 illustrates the specific conditions of a divergent shock compression using the example of a diverging spherical shock wave. Despite the spherical wave attenuation as it propagates, the values of the radial, σ_r, and circumferential, σ_θ, normal stresses directly behind the shock front lie on the same wave beam $\sigma_r = \sigma_\theta(1-v)/v$, as in the case of a plane wave. However, as the spherical flow expands, the ratio of stresses σ_r/σ_θ and their difference σ_r–σ_θ increase, as shown by the arrows in Fig. 6.13 *b*. As a result, the state of the material gradually approaches the fracture threshold. Thus, stress states that are close to the fracture threshold, unattainable in the case of a plane shock wave, can be realized behind the front of a diverging shock wave

The most natural and informative would be the use of diverging spherical or cylindrical shock waves to study the fracture of brittle

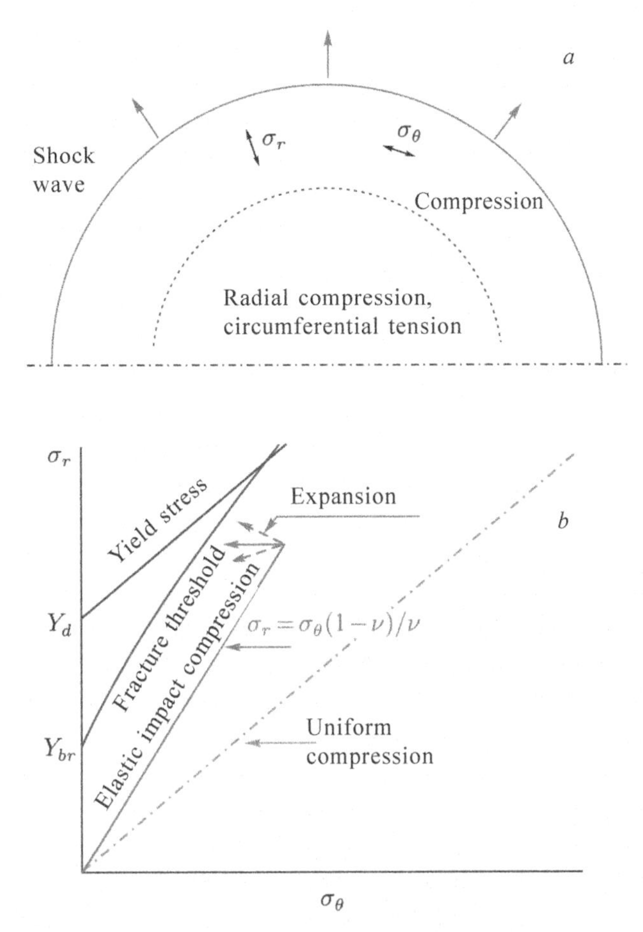

Fig. 6.13. Scheme (*a*) and diagram of possible trajectories of state change (*b*) under elastic compression of brittle material by a diverging spherical shock wave.

materials during their shock compression and to determine the thresholds of fracture and the conditions for the transition from fracture to plastic deformation. Experiments of this kind using specially organized detonation of explosive charges to generate spherical and cylindrical shock waves were carried out, but they turned out to be difficult to implement and not very informative. Under laboratory conditions, with reasonable sample sizes and explosive charges, it is quite difficult to implement loading conditions with divergent shock waves that would be close enough to the optimum for carrying out measurements.

In [64, 65], quasispherical diverging waves in plates of hard ceramics were created using weakly convex impactors. The advantages of this method are its relative simplicity and the ability

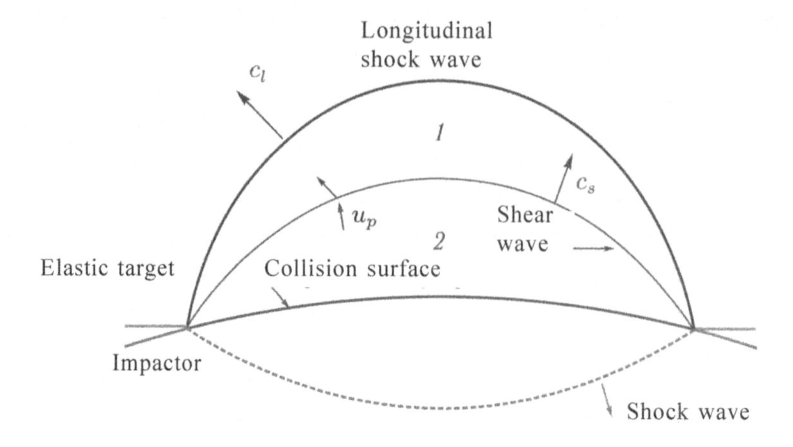

Fig. 6.14. The wave configuration realized at impact by a convex impactor on an elastic plane target.

to control the impact velocity and, accordingly, the intensity of the shock wave within wide limits. It can be shown that when a spherical impactor with a radius of curvature of the surface R_i and velocity u_i is struck, a diverging shock wave with a shape close to spherical and an initial radius of curvature equal to

$$R_S \approx R_i u_i / c_l,$$

where c_l is the longitudinal speed of sound in the target material. The task, however, is complicated by the fact that at stresses below the elastic limit in the target, in addition to the spherical compression longitudinal wave, a shear wave of a larger radius of curvature is generated, as shown in Fig. 6.14. The shear wave divides the flow into two parts: the divergent flow region 1 directly behind the diverging frontal shock wave region and the weakly divergent flow region 2. The divergence of the flow in region 1 can be approximately characterized by the radius of curvature of the longitudinal wave, which in the experiments discussed below was several centimeters. The divergence of the flow in region 2 is characterized by the curvature of the impact surface, which is several tens of centimeters. As a result, the diverging longitudinal wave attenuates rather rapidly as it propagates (the stress behind its front is inversely proportional to its radius), while in region 2 the stress drop as the distance from the impact surface occurs much slower.

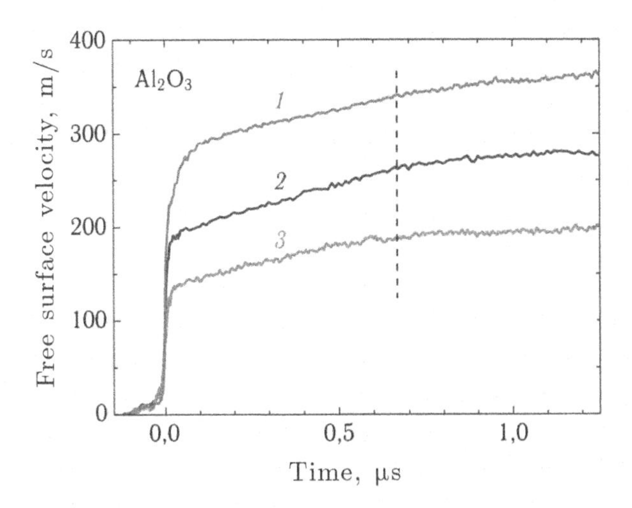

Fig. 6.15. The results of experiments with a divergent shock compression of Al_2O_3 ceramics.

Figure 6.15 shows the results of experiments with the divergent shock loading of ceramic alumina plates. In the experiments, the free surface velocity histories of copper reference plates with a thickness of 2.6 mm glued onto the back surface of a ceramic sample with a thickness of 10 mm were recorded. The use of a reference plate maintains high compressive stresses in the sample and prevents its fracture after the shock wave has reached the surface. In experiment 1, the radial stress at the entrance of the shock wave into the sample exceeded the Hugoniot elastic limit; in experiments 2 and 3 it was lower than this limit.

All wave profiles in Fig. 6.15 show a linear increase in the parameters behind the front of the longitudinal wave and a sharp decrease in the slope at the moment the shear wave reaches the surface. The wave profiles do not contain any obvious signs of fracture by compression. Nevertheless, the measurement results are suitable for estimating the range of stress states that are below the fracture threshold. For a simple estimate, we assume that the expansion of the spherical layer behind the front of a longitudinal elastic wave occurs at a constant average stress (pressure). In this case, it can be shown that

$$\dot{S}_\theta = -2G\frac{u_p}{r}, \quad \dot{S}_r = -2\dot{S}_\theta, \tag{6.7}$$

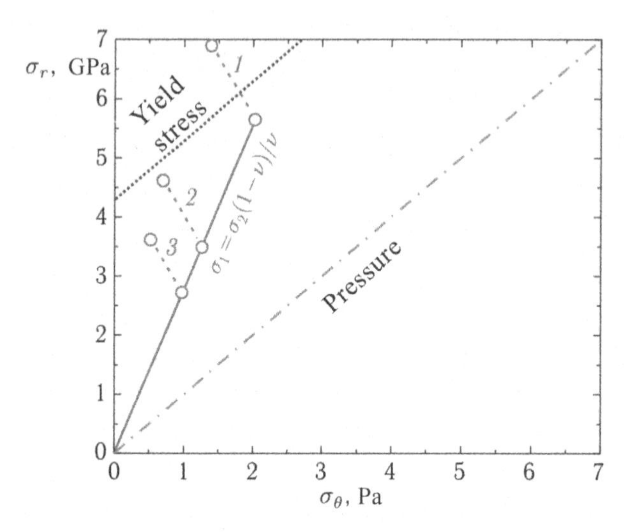

Fig. 6.16. Estimates of the stress states realized in experiments with a divergent shock loading of Al_2O_3 ceramics. The indices correspond to Fig. 6.15.

where S_r and S_θ are the radial and circumferential deviator stresses, respectively. Figure 6.16 shows the trajectories of the change in the stressed state thus estimated for the back layers of ceramic samples. The initial states on the trajectories correspond to the measured values of the surface velocity immediately behind the shock wave. The estimated spherical expansion time Δt = 0.65 μs corresponds to the time interval between the longitudinal and shear waves on the back surface of the sample. The flow radius r = 31–35 mm was estimated from the radius of curvature of the impactor and its velocity, taking into account the sample thickness. Thus, the obtained values of the stress state in experiment 1 go beyond the yield point. Consequently, despite the fact that the stress directly behind the shock wave front does not reach the dynamic yield strength, in the expansion process, plastic deformation of the outer layers of the sample begins. This circumstance may explain the smaller slope of the wave profile behind the shock wave front in this experiment than in the experiments 2 and 3. The estimates of the limiting states in the last two experiments do not contradict the data of static tests.

Figure 6.17 presents the results of similar experiments with B_4C ceramics. The Hugoniot elasticity limit of this ceramics is 14 GPa. The results of the experiment at a compression stress on the impact surface greater than this magnitude are qualitatively different from the wave profiles obtained at lower shock compression stresses:

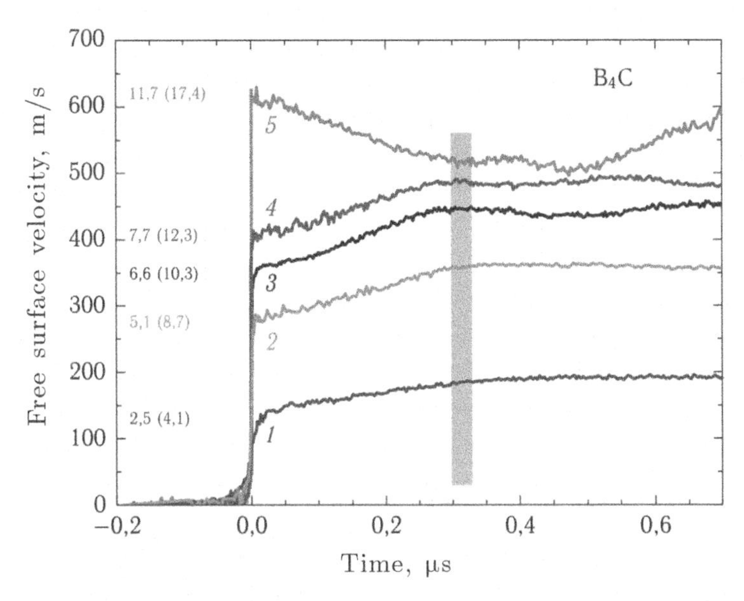

Fig. 6.17. The results of experiments with a divergent shock compression of B_4C ceramic plates with a thickness of 8 mm. The radial compression stresses are indicated, which correspond to the measured values of the velocity and (in brackets) the calculated values on the impact surface.

instead of an increase in velocity behind the front of the shock wave, in this experiment its fall is recorded. With a decrease in the impact velocity, the decrease in velocity behind the front (and, accordingly, the radial compression stress) is replaced by its increase. However, if in the case of aluminium oxide a practically linear growth of the parameters was recorded behind the shock front, for brittle ceramics, as was shown in the previous section, B_4C nonlinear growth is observed, which is probably due to the relatively slow fracture process in shock-compressed ceramics.

Thus, preliminary experimental studies show that divergent shock loading is a realistic method for varying the stress state of shock-compressed hard ceramic materials and determining the threshold conditions and patterns of their fracture.

6.5. Behaviour of sapphire under shock compression

Sapphire monocrystals are used as window material in space and defense technology, as well as in shock-wave experiments with optical recording. For these applications, it is important to ensure the preservation of the optical homogeneity of the material in a wide

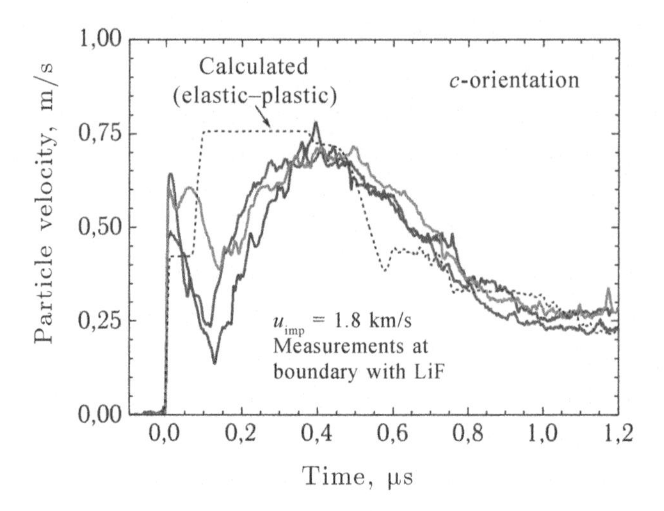

Fig. 6.18. The particle velocity histories measured at the interface between the sapphire sample of orientation *c* with a thickness of 5 mm and the LiF window at impact with a speed of 1.8 ± 0.05 km/s through an aluminium base plate 2 mm thick.

range of impact intensities. It is known that sapphire transparency is disrupted at shock compression stresses close to the Hugoniot elastic limit. In this case, the luminescence of shock-compressed sapphire is recorded, which has a pronounced heterogeneous character. Microscopic analysis of particles of sapphire samples after shock compression revealed a large number of dislocations and twins, indicating the implementation of plastic deformation mechanisms.

Figure 6.18 shows the results of three identical experiments [66] with sapphire samples with orientation *c* and a thickness of 5 mm at a speed of an aluminium impactor of 1.8 km/s. For comparison, the particle velocity history obtained as a result of computer simulation in the framework of the approximation of an ideal elastoplastic body with zero tensile strength for the same loading conditions is also given. The measured wave profiles are qualitatively different from those recorded for ceramic alumina. The wave profiles in Fig. 6.18 are poorly reproduced from shot to shot and contain irregular oscillations, while for ceramics the wave profiles are regular and well reproducible. Strong oscillations and low reproducibility of wave profiles for sapphire are obviously the result of considerable heterogeneity of inelastic deformation. For ceramics, a monotonous increase in the parameters behind the elastic precursor front with a smooth transition to a plastic compression wave is recorded, while in the case of sapphire there is a significant decrease in the parameters behind the elastic precursor front.

From a comparison of the wavelengths measured and calculated assuming zero tensile strength of the wave profiles in Fig. 6.18 in the final part of unloading ($t > 0.8$ μs), it follows that after inelastic deformation in the shock wave the sapphire strength becomes negligible. Otherwise, there is a deeper drop in the particle velocity in the rarefaction wave at the boundary with a low-impedance window. An analysis by the method of the characteristics of wave interactions upon reflection of a compression pulse from a lithium fluoride window shows that the recorded decrease in velocity behind the peak corresponds to the values of tensile stresses in the reflected wave to –(2–2.5) GPa. Since after inelastic deformation the sapphire loses tensile strength, the preservation of high strength immediately behind the precursor front can be considered as evidence of the purely elastic nature of deformation in this part of the compression pulse.

The wave profiles measured in experiments with samples of orientations *r*, *d*, *n*, and *g* are generally similar to those discussed above and have no new features. More interesting are the results of experiments with the sapphire orientations *s* and *m*. Figure 6.19 compares the wave profiles for the sapphire samples of these orientations at an intermediate impact velocity. There are some irregular oscillations on the profiles, but they have a much higher frequency and a smaller amplitude than in the experiments with the

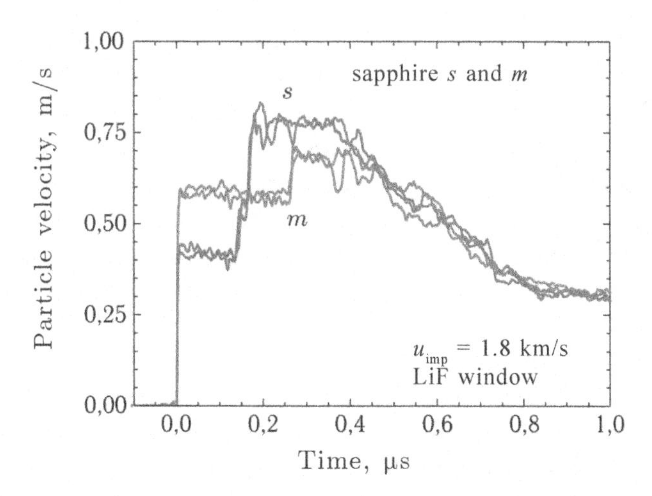

Fig. 6.19. The results of two pairs of experiments with sapphire samples of orientation *s* and *m* 5 mm thick at an impact velocity of 1.8 km/s.

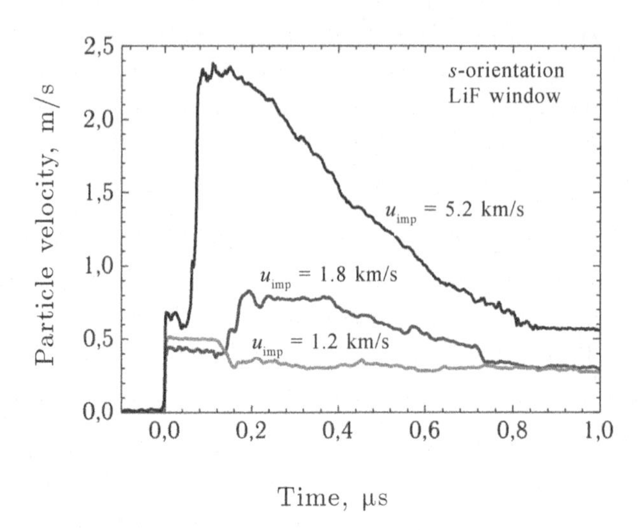

Fig. 6.20. Results of experiments with sapphire with orientation s at three impact velocities (shown at the wave profiles).

other orientations listed. The speed of the second (plastic) wave at an intermediate impact velocity is 8.3 km/s, which is reasonably consistent with the sapphire bulk modulus. The rise time of the parameters in a plastic wave is not more than 30 ns, which is five times less than in the sapphire with orientation *c*. The stepped form of a plastic wave is obviously the result of multiple reflections of an elastic wave between a window and a plastic shock wave. A short rise time and a high frequency of noise oscillations indicate a short stress relaxation time and a relatively high degree of plastic deformation homogeneity under the shock compression of the sapphire of these two orientations.

Comparison of measurement results at different impact speeds in Fig. 6.20 demonstrates the anomalously nonmonotonic dependence of the stresses behind the precursor front on the final shock compression pressure. Usually, in the experiments with shock waves in solids, the decay of elastic precursors as they propagate, caused by the onset of plastic deformation and the corresponding stress relaxation behind its front, is observed. In this case, the initial stress in the precursor in the immediate vicinity of the impact surface increases with increase of the pressure of shock compression. As a result, at a fixed distance travelled by the wave, the stress behind the precursor front is also higher, the higher the shock pressure is.

The behaviour of sapphire under shock-wave loading is associated with a number of specific features, some of which were not observed for other materials. Irregular stress oscillations, which are most pronounced in the shock compression of sapphire in the direction of the *c* axis, often accompany the formation of twins. Such oscillations in twinning appear on the stress–strain diagrams of metals and alloys. Unusual and unexpected is the fact that for several sapphire orientations, the compression stress behind the elastic wave front with an increase in impact velocity from 1.2 km/s to 1.8 km/s does not increase and does not remain the same, as it should be for an elastic–plastic material but decreases. This is probably due to the fact that for the nucleation of twins in a crystal, much higher stresses are required than for their growth. In this case, the emergence of twins under the action of an applied stress occurs, judging from the reproducibility of the wave profiles, randomly, and the rate of nucleation increases with increasing applied stress. The recorded value of the Hugoniot elastic limit of sapphire depending on the direction of compression and the impact velocity varies from 12.4 GPa to 24.2 GPa. The highest values of the elastic limit are recorded under uniaxial shock compression in the *c* direction perpendicular to the base plane of the crystal and in the *m* direction perpendicular to the prismatic plane. Compression in the *c* and *m* directions eliminates the creation of shear stresses in the base plane and, accordingly, the activation and movement of dislocations and twins in this plane. When compressed in the direction of the *c* axis of the crystal, the

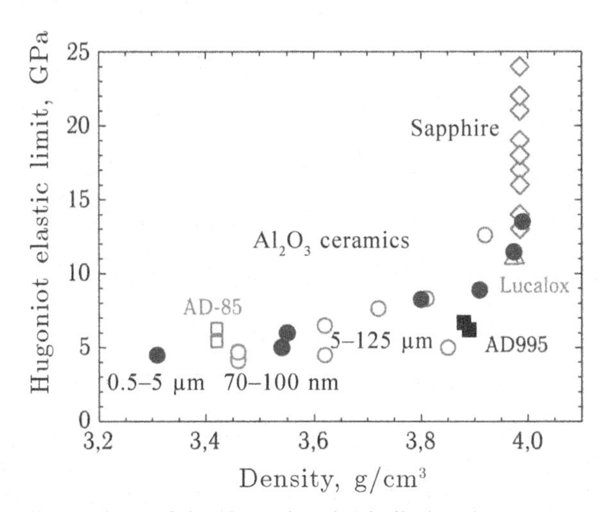

Fig. 6.21. Comparison of the Hugoniot elastic limit values measured for sapphire of different orientations and for several grades of ceramic alumina of different density.

mechanisms of sliding and twinning in prismatic planes are also excluded. The apparent difference in the wave profiles of the shock compression of sapphire in the directions *s* and *m* compared with the data for other directions indicates with a high probability the prevailing contribution of the dislocation mechanisms of plasticity with such crystal orientations relative to the load.

In Fig. 6.21 the results of measurements of the Hugoniot elastic limit of the sapphire of different orientations are compared with those for ceramic alumina of various densities. It can be seen that the elastic limit of the hardest hot-pressed ceramics with a density practically coinciding with the sapphire density does not exceed the minimum values of the sapphire elastic limit obtained for orientations *r* and *n*.

7

Investigations of polymorphic transformations and phase transitions at shock compression

The development of high static pressure physics methods, where it is possible to use fine modern diagnostics requiring a large exposure, has made dynamic measurements less competitive in terms of constructing phase diagrams of a substance. Apparently, the most interesting fundamental problem of polymorphism of solids under shock compression is the question of the mechanism and kinetics of high-speed transformations. In this regard, studies of the behaviour of materials in different initial structural states and in a wide temperature range are carried out.

7.1. Polymorphic transformation of iron under pressure

Typical examples of measurement results are shown in Fig. 7.1, which shows the free surface velocity histories of Armco iron samples loaded with an impact of an aluminium plate at temperatures from 20°C to 600°C, and pressure histories measured in internal sections of the samples. The transformation of the α-phase of iron with a BCC crystal structure into a high-pressure ε-phase with a hexagonal close-packed structure is interesting because it was first discovered namely in experiments with shock waves.

On the velocity histories, the exit to the surface of the sample of three successive compression waves is recorded. Due to the increase in the longitudinal compressibility with the transition from elastic

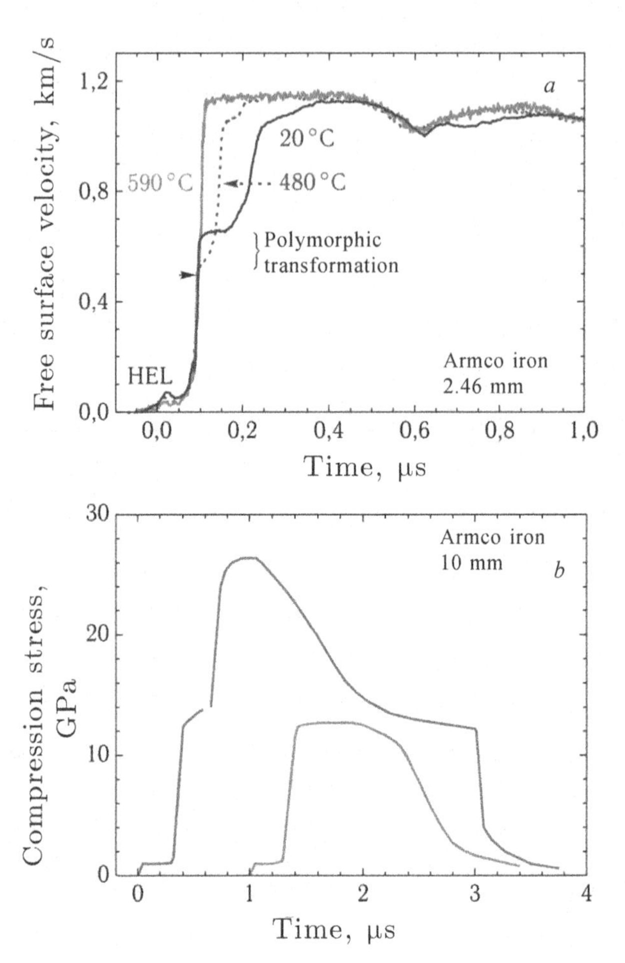

Fig. 7.1. Examples of measurement results of wave profiles of pressure and velocity. *a*: free surface velocity histories of plane samples of Armco iron 2.46 mm thick after impact by an aluminium plate 2 mm thick with a speed of 1.9 ± 0.05 km/s. The measurements were carried out at normal and elevated temperatures (indicated near the corresponding curves). The pressure of shock compression of iron was 19 GPa. *b*: Stress profiles in shock compression pulses generated in Armco iron samples by the impact of aluminium plates with a speed of 1.05 km/s and 2.06 km/s. Measurements by the method of manganin pressure gauges.

to plastic deformation, the shock wave loses its stability and splits into an elastic precursor and the next plastic compression wave. At a pressure of ~13 GPa, iron undergoes an $\alpha \rightarrow \varepsilon$ polymorphic transformation (BCC → HCP) with a decrease in specific volume, as a result of which the plastic compression wave in this pressure range splits into two. The pressure behind the front of the first plastic shock wave corresponds to the beginning of the transformation,

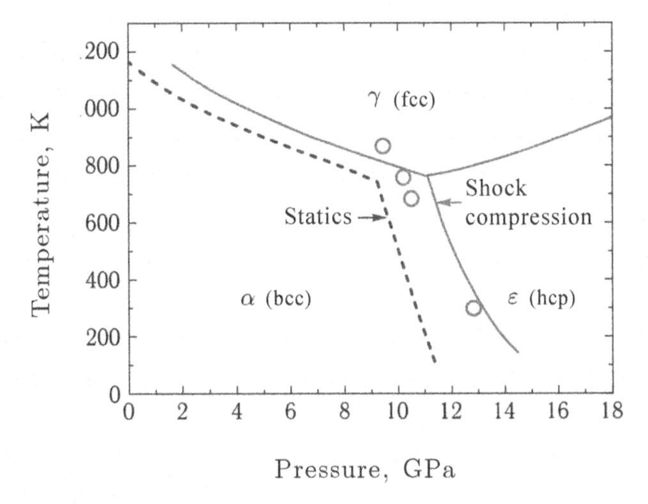

Fig. 7.2. The iron phase diagram. The dots show the results of processing the free-surface velocity profiles shown in Fig. 7.1.

and its decay and compression rate in the second plastic wave are determined by the kinetics of the structural transformation. Figure 7.1 *b* shows the complete pressure profiles in iron, obtained with manganin gauges. In these experiments, it was possible for the first time to observe directly the rarefaction shock wave formed as a result of the reverse structural transformation of iron into the low-pressure phase.

Reducing the time interval between the waves at elevated temperatures is mainly the result of a decrease in the transformation pressure (and, therefore, the speed of the first plastic shock wave) and a corresponding increase in the pressure increment in the second plastic wave (and, therefore, its speed). Figure 7.2 summarizes and compares with literature data the results of pressure measurements of the $\alpha \rightarrow \varepsilon$ transformation at various test temperatures. Since the almost linear temperature dependence of the transformation pressure is preserved in the temperature range studied by us, it is likely that $\alpha \rightarrow \varepsilon$ transformation occurs in the entire studied range, and the transition to the FCC structure occurs at temperatures above 590°C.

Judging by the steepness of the compression waves in Fig. 7.1, the $\alpha \rightarrow \varepsilon$ transformation time in iron decreases with increasing temperature. This reduction is partly due to an increase in the amplitude of the second plastic compression wave as the pressure of the onset of transformation decreases. More systematic studies of the effect of temperature on the rate of polymorphic transformation under shock compression, carried out for potassium chloride crystals [67], have shown, against expectations, some slowdown in the transformation with heating. Since, along with the slowing down of the transformation, a similar slowing down of the shear stress relaxation was also observed in the elastic precursor, the authors of [67] concluded that the dislocation mechanism of the structural rearrangement of shock-compressed KCl crystals was involved.

7.2. The transformation of graphite into diamond under shock compression

Due to practical importance, considerable attention was paid to measuring the parameters of the transition of graphite to diamond. The transformation of graphite into diamond under shock compression conditions attracts attention in connection with the known and potential applications. It is known that the equilibrium pressure for the conversion of graphite to diamond increases with heating from 1.7 GPa at 0 K to 12 GPa at a temperature of 5000 K at the graphite–diamond–liquid carbon triple point [68]. Industrial synthesis of diamond is carried out using catalysts (transition metals) in the pressure range from 5 to 12 GPa and temperatures of 2000–3000 K [69, 70], close to the equilibrium line between graphite and diamond. At the same time, the pressure of direct conversion of graphite to diamond at temperatures of 300–1000 K is approximately 20 GPa, both under the static compression conditions, where the process takes minutes or tens of minutes, and under the shock compression in the submicrosecond range of duration [70]. According to the measurement results of shock compressibility, the density of compressed graphite after transformation is close to the density of diamond at the corresponding pressures and temperatures. According to static measurements, the high-pressure phase is characterized by low electrical conductivity and significant transparency. Optical studies did not reveal any signs of crystalline diamond formation up to 55 GPa in the Raman spectra of compressed graphite, but X-ray structural measurements under pressure showed that the high-pressure phase is a hexagonal shape of a diamond (londsdaleite). The correlation of the orientations of the crystallographic planes indicates the martensitic nature of the transformation of graphite into hexagonal diamond under pressure at room temperature. At the same time, however, the high pressure phase during unloading is not preserved or almost not preserved. Annealing under pressure at a temperature of 1000°C makes it possible to partially preserve hexagonal diamond [71].

The material stored after annealing under pressure of 12–25 GPa in the range of higher temperatures of 1000°C–2000°C is an opaque mixture of graphite, londsdaleite and cubic diamond. With a further increase in temperature, the material obtained becomes a transparent polycrystalline cubic diamond. With an increase in the annealing temperature under pressure the grain size of polycrystalline diamond increases from 4–10 nm at 1500°C to 10–30 nm at 2300°C and 30–200 nm at 2600°C. With an increase in the annealing temperature, the contribution of the martensitic transformation mechanisms decreases. Under the shock compression conditions, the conversion of graphite to diamond or a diamond-like high-pressure phase can occur in a time of about 10^{-8} s, which is usually explained by the deformation (martensitic) nature of the structural rearrangement mechanism under these conditions.

Figure 7.3 presents the results of experiments [72] with samples of highly oriented pyrolytic graphite (HOPG) with different misorientation angles of the mosaic structure, in comparison with the data for pressed highly ordered graphite OSCh-T1. In all cases, the splitting of the shock wave with the formation of a two-wave

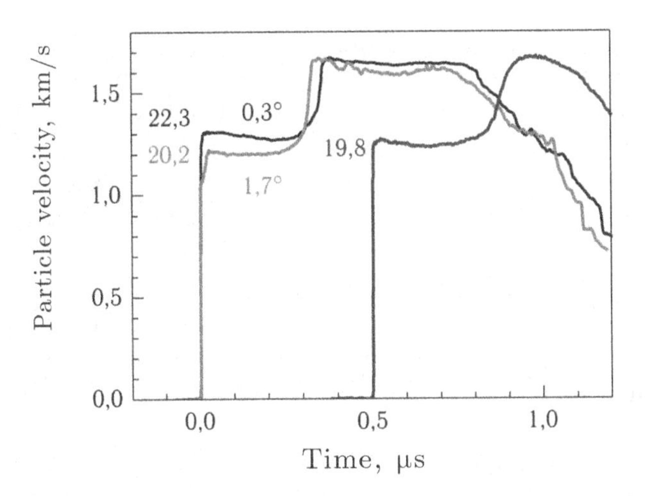

Fig. 7.3. Wave profiles of shock compression of highly oriented pyrolytic graphite (HOPG) of two types with different mosaic scattering angles in comparison with the data for OSChT-1 pressed natural graphite with a density of 2.17 g/cm³. The thickness of HOPG samples with mosaic scattering angles of 0.3° and 1.7° is 1.7 mm and 2.1 mm, respectively; sample thickness of compressed graphite OSChT-1 is 4.25 mm..

structure is clearly recorded, which is a consequence of the sharp increase in compressibility during the polymorphic transformation. The parameters behind the front of the first shock wave correspond to the onset of the transformation of graphite into a denser phase at high pressure. It can be seen that, against expectations, the pressure of transformation of HOPG with a large mosaic-scattering angle is 10% lower than in more perfect graphite with a small misorientation of crystallites. The temperature behind the front of the first shock wave in HOPG in accordance with the equation of state of graphite of maximum density is 290°C at a transformation pressure of p_t = 20.2 GPa in the case of HOPG with mosaic scattering of 1.7° and 330°C at p_t = 22.3 GPa in the case of HOPG with a misorientation of 0.3°. The presence of a small porosity in pressed graphite leads to a greater energy deposition by shock compression, with the result that the temperature behind the front of the first shock wave in OSChT-1-1 is 410°C at a pressure of 19.8 GPa. For comparison, let us point out that the pressure of transformation in the widespread pyrolytic graphite UPV-1 with a low degree of three-dimensional ordering exceeds 30 GPa.

The steepness of the second compression wave is proportional to the rate of transformation into the high pressure phase. The comparison of the wave profiles in Fig. 7.3 shows that the rate of transformation of the two types of HOPG is higher than in the case of the pressed highly ordered graphite. While in experiments with the HOPG the rate of transformation increases with compression and remains high until it ends, in the case of pressed graphite, the transformation process at the final stage slows down.

A relatively slow drop in particle velocity and, accordingly, pressure behind the front of the first shock wave can have two probable causes. When the first shock wave is reflected from a more rigid window, the pressure in graphite increases by about 10–20%, as a result of which its transformation into a dense phase is initiated (or accelerated), which in turn leads to pressure relaxation. On the other hand, such qualitative features of the wave profiles can be associated with the kinetics of transformation, namely, with its acceleration as it develops.

Figures 7.4 and 7.5 show the results of experiments with graphite of two brands with different orientations of the basal planes relative to the direction of shock compression. In the case of uniaxial shock compression in the transverse direction (the plane of the shock front is perpendicular to the basal planes), the recorded pressure behind

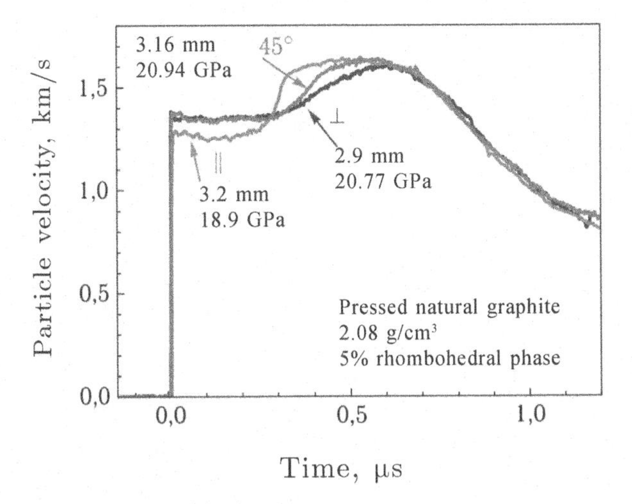

Fig. 7.4. Particle velocity histories at the interface between graphite and a lithium fluoride window for fine-grained natural graphite with a density of 2.08 g/cm^3 and a rhombohedral phase content of 5% with orientation of the basal planes parallel (||), perpendicular (⊥) and inclined at an angle of 45° to the plane shock wave front. The sample thicknesses and pressure values behind the first shock wave are indicated.

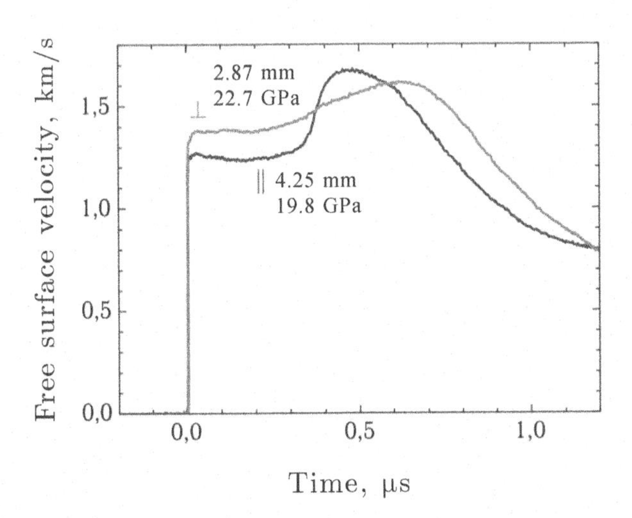

Fig. 7.5. Particle velocity histories at the graphite–LiF contact interface for the OSCh-T1 graphite with orientation of the basal planes parallel (||) and perpendicular (⊥) to the plane of the shock wave front. The sample thicknesses and pressure values behind the first shock wave are indicated.

the first shock wave is 5–10% higher and the rate of increase of the parameters in the second compression wave is several times smaller than during compression in the *c* axis direction of graphite crystals (when the front plane of the shock wave is parallel to the basal planes). The difference is more pronounced in the case of more ordered graphite. Measurements with the orientation of shock compression at an angle of 45° to the basal planes of graphite showed the same increased pressure of transformation as under compression in the transverse direction, but a higher rate of transformation in the second wave.

7.3. Search for structural transformation under tension

Extremely promising is the search for structural transformations under tension, similar to those occurring under compression. In addition to the natural hope of obtaining materials with unusual properties, this search is stimulated by the problems of strength of materials. Some calculations do predict structural transformations in the domain of negative pressures, in particular, graphitization of the diamond at stretching, transition to the clathrate phase in silicon. The results of ab initio calculations of the zero-temperature iron isotherm [33], shown in Fig. 7.6, demonstrate a volume jump associated with the rearrangement of the energy spectrum of a crystal at a pressure of –3.4 GPa. The presence of a site of anomalous compressibility in the vicinity of −3.4 GPa should lead to the formation of shock jumps

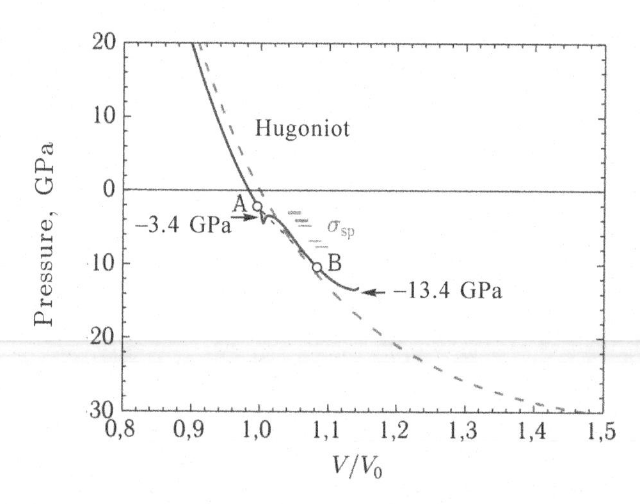

Fig. 7.6. Calculated zero isotherm (solid line) and extrapolated Hugoniot of α-iron (dashed line). The dashes show the measured values of fracture stresses (spall strength) of iron for various durations of shock-wave load. AB is the region of anomalous compressibility where the formation of a shock rarefaction wave was expected.

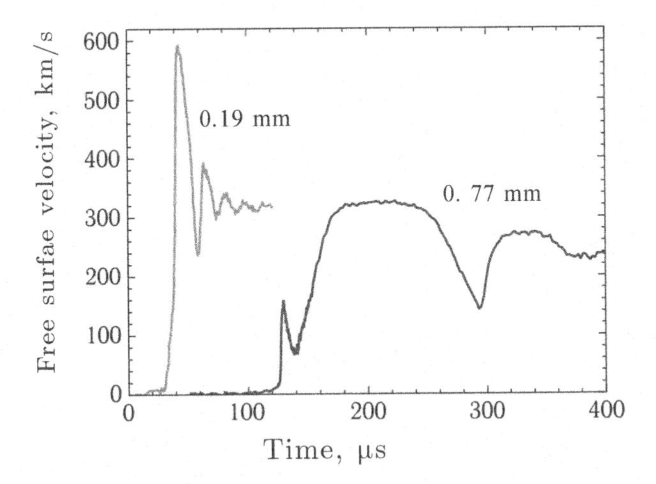

Fig. 7.7. The free surface velocity histories of high-purity iron plates with a thickness of 0.19 and 0.77 mm when struck with aluminium plates with a thickness of 0.05 mm and 0.4 mm with a speed of 1200 m/s and 650 m/s, respectively.

during the propagation of rarefaction waves in iron, which makes it possible to experimentally verify the presence of this anomaly. For this purpose, a series of experiments with high-purity iron single crystals was carried out. Figure 7.7 shows examples of the measured free surface velocity histories of single-crystal iron samples with a tenfold difference in the duration of the shock load. The wave profiles are generally similar to those obtained for other metals and, unfortunately, do not contain any evidence of the formation of shock rarefaction waves.

As shown in Fig. 7.6, the range of tensile stresses achieved significantly overlaps the region of the anomaly on the calculated zero isotherm of iron. Since electronic topological transitions should be practically inertia-free, it seems unlikely that the compressibility anomaly associated with them did not manifest itself due to the short duration of the effect of negative pressures. Perhaps the iron compressibility anomaly occurs only at lower temperatures and disappears with heating.

In this case, the instability of the crystal structure may be one of the factors determining the phenomenon of cold brittleness of iron.

The literature discusses the possibility of transformation of austenitic steel into martensite under the influence of negative pressures. Figure 7.8 shows the results of experiments [73] with samples of 12Cr18Ni10Ti stainless steel. As delivered, this steel is usually in the austenitic γ-phase (face-centered cubic lattice);

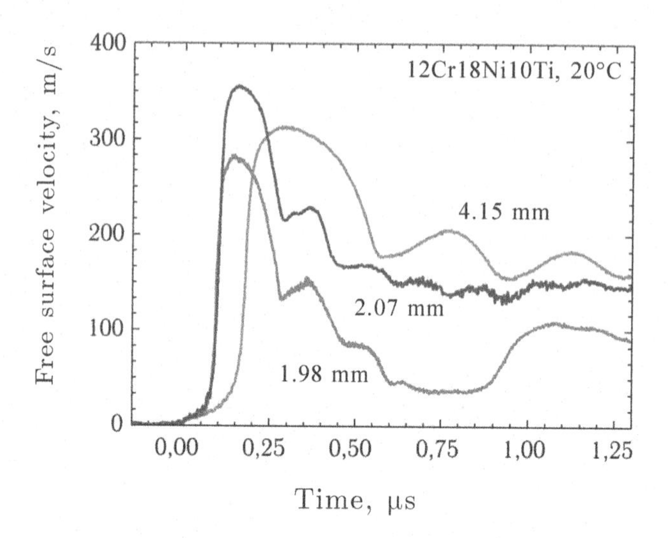

Fig. 7.8. The free surface velocity histories of samples of homogenized austenitic steel 12Cr18Ni10Ti with a thickness of 2 mm and 4 mm after being hit by a copper flyer plate with a thickness of 0.4 mm or 0.8 mm with a speed of 300 ± 50 m/s.

plastic flow at a temperature not higher than 100°C induces a phase transition of austenite into α-martensite (body-centered cubic lattice) in it. The phase formed in this way is called strain-induced martensite. The material used in the experiments was subjected to homogenizing heat treatment.

The most interesting feature of behaviour of the steel in Fig. 7.8 is the very prolonged nature of the spall fracture: the spalled plate retains its connection with the main sample and decelerates for a long time after the onset of fracture. Figure 7.9 presents the results of experiments with thick impactors, in which the growth of tensile stresses before spalling occurs more slowly, and the fracture zone itself is located farther from the back surface of the sample than in experiments with thin flyer plates. The most interesting and unexpected feature of the material behaviour under these loading conditions is the anomalous dependence of the velocity pullback before spalling Δu_{fs} on the duration of the rectangular load pulse.

Usually, with an increase in the ratio of the thickness of the impactor and the sample, the recorded rate decrement Δu_{fs} decreases, which is explained, firstly, by the fact that the local strain rate decreases before fracture, and secondly by the fact that the correction δ increases in relation (3.47), which characterizes distortion of the velocity profile due to the difference in the propagation velocities

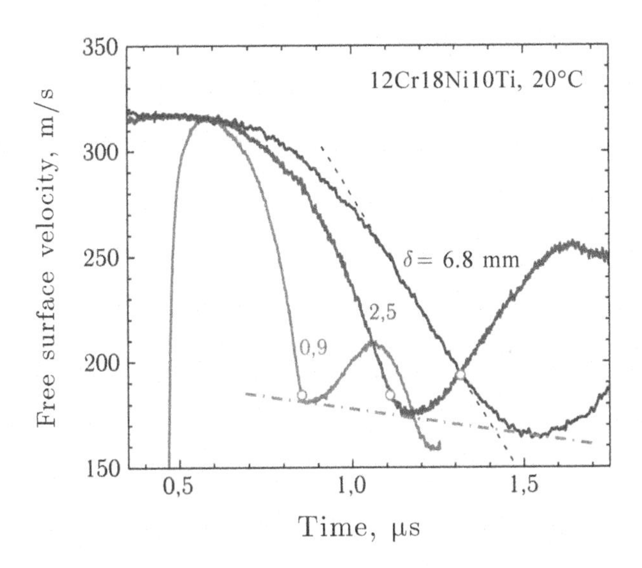

Fig. 7.9. The free surface velocity histories of steel samples with a thickness of 2 mm, 4 mm and 8 mm after the impact of a copper plate with a thickness of 0.8 mm, 2 mm or 5.5 mm, respectively. The values of the spall thicknesses are indicated. The wave profiles are shifted relative to each other in such a way as to clearly demonstrate the anomalies in dependence of the velocity pullback Δu_{fs} on the duration of a rectangular load pulse.

of the front of the spall pulse and the plastic part of the incident unloading wave ahead of it. If the spall fracture stress is not associated with the minima on the free surface velocity profiles, but with the onset of a deviation from the wave profile of the incident compression pulse (indicated by round markers on the wave profiles), then the anomaly in the dependence of the spall fracture parameters on the strain rate disappears.

Under normal conditions, strain-induced martensite is formed in places of large plastic strains; during spalling, in the vicinity of the pores. Since the martensitic phase is characterized by a higher yield strength, the formation of strain-induced martensite probably determines the prolonged spall fracture of steel. On the other hand, the relative slow martensitic transformation under the action of negative pressures could explain the appearance of an anomaly in the dependence of the spall strength on the strain rate shown in Fig. 7.9.

7.4. Melting under shock compression and tension

Figure 7.10 schematically shows the phase diagram of a substance,

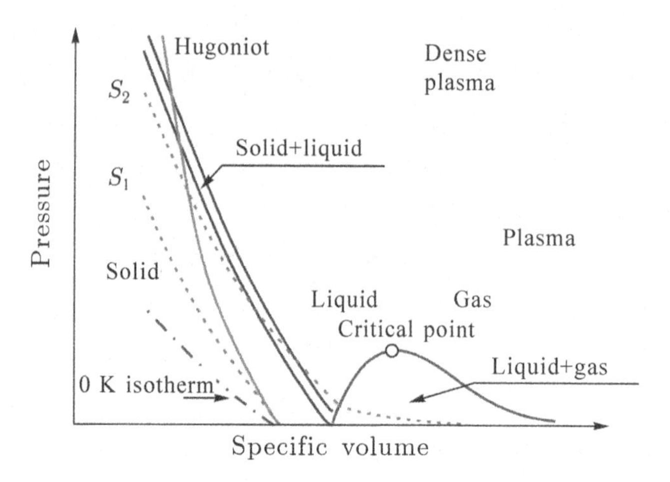

Fig. 7.10. The phase diagram of the substance.

including melting, evaporation, condensation, the critical point and the areas of the mixed solid and liquid phases and the liquid and gaseous phases. The melting point, as a rule (although there are exceptions), increases with pressure. The ratio of the slopes of isentropes, Hugoniots, and the melting curve is such that when heated in the process of isentropic compression, melting is not achieved, but with shock compression, the temperature increase is significant, and at some pressure, the Hugoniot should intersect the melting curve. In addition, melting can occur during isentropic unloading from a shock-compressed state; evaporation may occur at higher parameters of shock compression in the unloading substance. The parameters of melting and evaporation are important for a number of practical applications, in particular, for the design of protective anti-meteorite shields of spacecraft, and are subject to experimental research.

The question of the position of the solid–melt equilibrium line in the megabar pressure range remained unresolved for a long time. Under the conditions of static compression, the melting curves $T_m(p)$ of metals were constructed based on the results of measuring the jump of resistance at melting but the range of the parameters was methodically limited to pressures of 40–70 kbar (4–7 GPa), which is of course far from the megabar range. It was possible to advance in this direction in the 1980s [74] by means of mechanics, based on the obvious fact that the yield strength of the material should vanish when melting. Accordingly, the elastic part of the unloading wave propagating through the shock-compressed material, whose

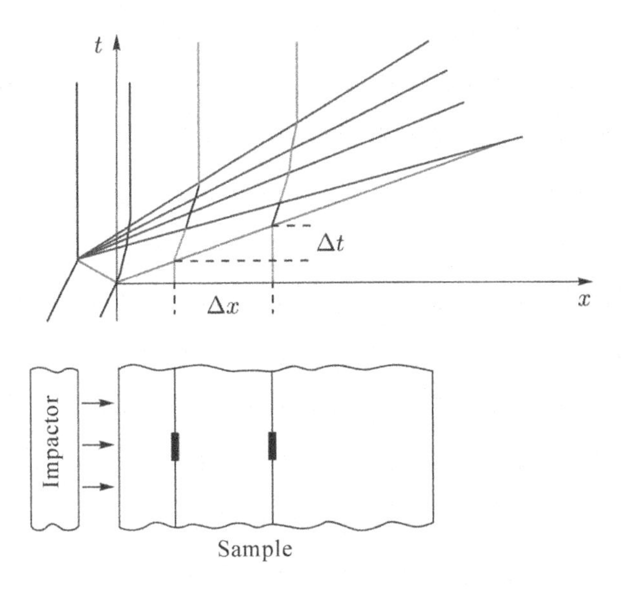

Fig. 7.11. The evolution of a compression pulse generated by a plate impact.

temperature exceeded the melting point as a result of irreversible heating, should disappear. As a result, the velocity of the rarefaction wave front which for a solid is equal to the longitudinal speed of sound at the corresponding pressure decreases to the value of the bulk sound speed. Methods for measuring sound speeds in shock-wave experiments are illustrated by the distance x – time t diagram in Fig. 7.11.

The loading of the sample is carried out by the impact of the plate. In the most illustrative case, the stress profiles $\sigma_x(t)$ are simultaneously recorded in two sections of the sample, an example of which is shown in Fig. 7.12. Knowing the distances between the gauges and determining the time intervals between the arrival times of the shock and rarefaction fronts on the first and second gauges using the experimental oscillogram, it is easy to find the velocity of the shock wave and the velocity of the rarefaction wave front propagating through the compressed substance. If the values of the velocity of the shock wave and the thickness of the impactor at the time of the impact are precisely known from independent measurements, then to determine the unloading front speed a one profile $\sigma_x(t)$ is sufficient. Finally, the magnitude of the velocity of the rarefaction wave front can be found by measuring the distance at which the decay of the shock wave begins.

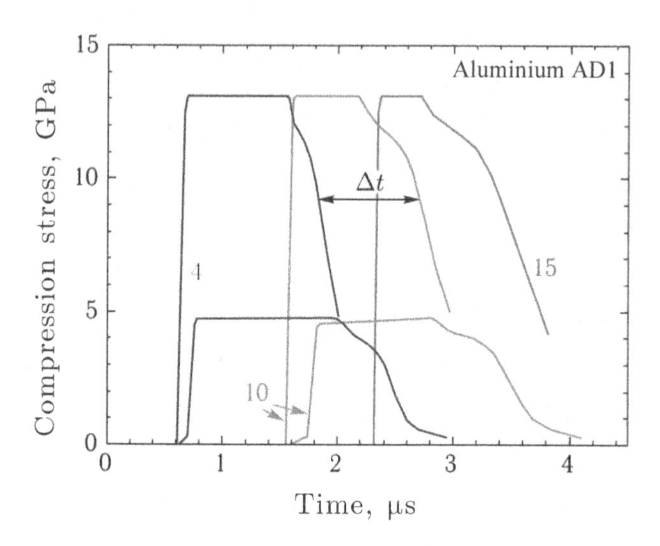

Fig. 7.12. The propagation of elastoplastic compression and rarefaction waves in aluminium after a plate strike at different speeds. Measurements with manganin pressure sensors.

In the latter case, the application of the optical method of indicator barriers [74] turned out to be particularly fruitful, where the luminescence of the shock wave in a thick layer of an organic liquid or other transparent material placed directly behind the plane sample of the material under investigation is recorded. Experiments are carried out with stepped-shaped samples, which makes it possible to measure with several ratios of the thickness of the impactor and the sample in one experiment. The intensity of the radiation emitted by the shock-compressed substance of the barrier indicator is constant until the rarefaction wave catches up with the shock wave in the barrier, after which the radiation intensity begins to decrease. The duration of the time interval between the moments of the appearance and the beginning of a decrease in the intensity of radiation is a linearly decreasing function of the sample thickness; extrapolation of this function to zero duration gives the thickness at which the rarefaction wave catches up with the shock wave just on the sample surface. In the case of a symmetric impact, the determination of the velocity of the rarefaction wave front in a shock-compressed sample then does not represent difficulties. The method turned out to be very accurate and especially effective in the field of high pressure shock compression.

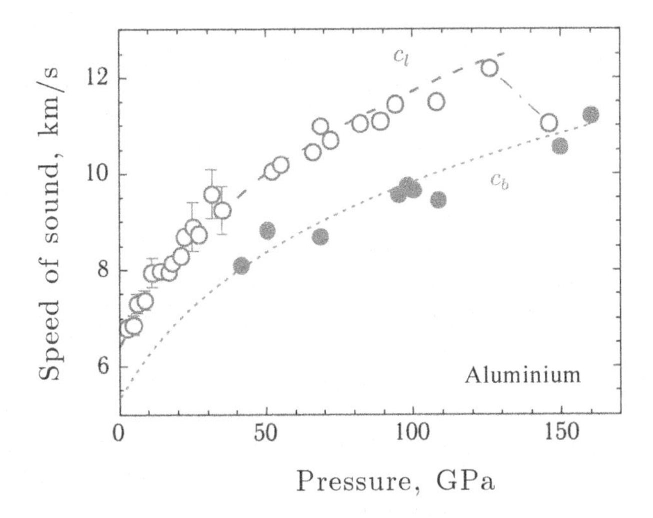

Fig. 7.13. The dependence of the velocity of the front of the rarefaction wave and its middle part in aluminium on pressure. Points are experimental data; the lines indicate the estimates of the bulk speed of sound in the quasi-acoustic approximation and the longitudinal speed of sound in the approximation of the constant Poisson ratio.

Figure 7.13 shows the results of measuring the velocity of the front and middle parts of the rarefaction waves in aluminium in a wide pressure range. The middle part of the rarefaction wave propagates with a bulk speed of sound, which, as it turned out, is calculated with good accuracy from the condition of coincidence of the Hugoniot and Riemannian isentrope in the pressure–particle velocity coordinates (see Ch. 1). In the Lagrange coordinates, the bulk speed of sound a is calculated in this approximation as

$$a = c_0 + 2bu_p \quad \text{or} \quad a = \sqrt{c_0^2 + 4bp/\rho_0},$$

where c_0 and b are the coefficients of the linear expression for the Hugoniot connecting the velocity of the shock wave and the particle velocity.

The values of the longitudinal speed of sound with good accuracy are described under the assumption that the Poisson's ratio is constant, which corresponds to the constancy of the ratio of the values of the longitudinal and bulk speeds of sound. In the case of iron, steel, and other materials that undergo rearrangement of the crystal structure during compression, the Poisson's ratio is not

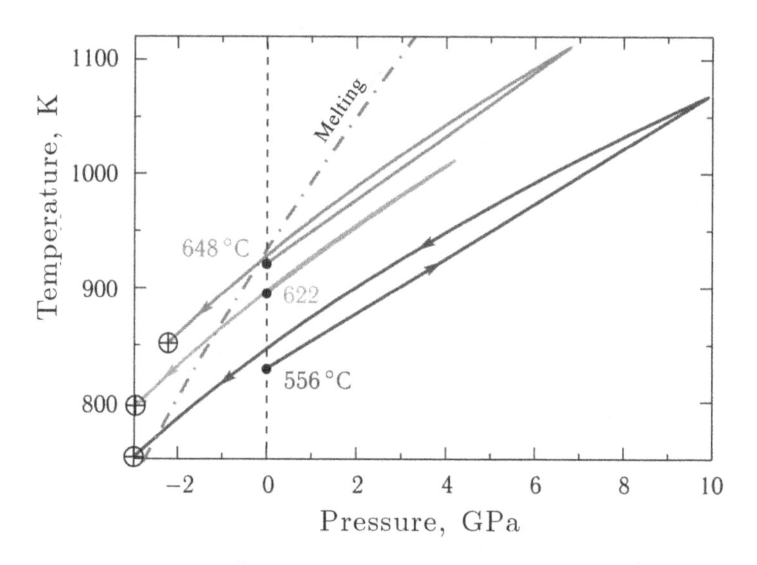

Fig. 7.14. Changes in the state of aluminium single crystals under shock compression and subsequent tension for the experimental conditions shown in Fig. 4.23.

constant, and such estimates are applicable only for the pressure region below the polymorphic transformation.

From Fig. 7.13 it is clear that the calculated dependences agree satisfactorily with experimental data at shock compression pressures up to 125 GPa, above which the measured velocity of the rarefaction front in aluminium decreases and at p = 150 GPa becomes equal to the bulk speed of sound. The disappearance of the difference between the velocity of the rarefaction wave and the bulk speed of sound is explained by the melting of aluminium in the shock wave. The method of determining the melting point on the Hugoniot by comparing the velocity of the rarefaction wave front and the bulk speed of sound has been successfully used for a wide class of materials.

Measurements of the spall strength of metals at elevated temperatures raised the question of the possibility of melting in the area of negative pressures. Figure 7.14 shows part of the phase diagram of aluminium, calculated for the conditions of spall experiments with single crystals [17], the results of which are shown in Fig. 4.23. The position of the boundaries of the coexistence region of solid and liquid phases at negative pressures is determined, as in the compression region, by the equality of the chemical potential of the phases. In this sense, the boundary of the melting region, including that part of it which is in the region of negative pressures, is equilibrium.

Since all states with negative pressures are metastable, melting under tension, if one can be observed, is the transformation of a metastable solid phase into a metastable liquid.

The onset of melting should be accompanied by an increase in compressibility and a decrease in the flow stress, which should cause the appearance of anomalies in the rarefaction wave profile with an entrance to the two-phase region. However, the wave profiles near the melting point are fully similar to those measured at lower temperatures. The intersection of the calculated boundary of the melting region under tension is not accompanied by a sharp drop in the strength of single crystals. It follows that in the experiments performed, the material did not melt, and the measured strength in all cases corresponds to the strength of the solid.

If the expected melting in the process of high-rate tensile loading at high temperatures did not occur, then, consequently, in experiments with single crystals superheated solid states were realized. The magnitude of the overheating reached 60–65°C with the shortest duration of the shock load. It is believed that the crystal surface plays a critical role in melting, where the activation energy is close to zero. The melting of a uniformly heated solid always begins with its surface. The superheated solid states can be created only inside the body provided that its surface has a temperature below the melting point. This condition was implemented in the shock wave experiments.

7.5. Solidification in dynamic compression

Solidification under dynamic compression is not so important practically; nevertheless, it also attracts the attention of researchers.

In particular, this applies to water, which in the solid state can exist in several crystalline modifications. Figure 7.15 shows a part of the phase diagram of water from which it follows that the irreversible heating of a substance in a shock wave does not allow the possibility of the transformation of water into ice with a single shock compression. However, this transformation is possible with shockless isentropic and stepwise shock compression.

In the experiments described below, using a stepwise shock compression, water was poured into a sealed cell, the bottom of which was a metal base plate, through which a shock wave was

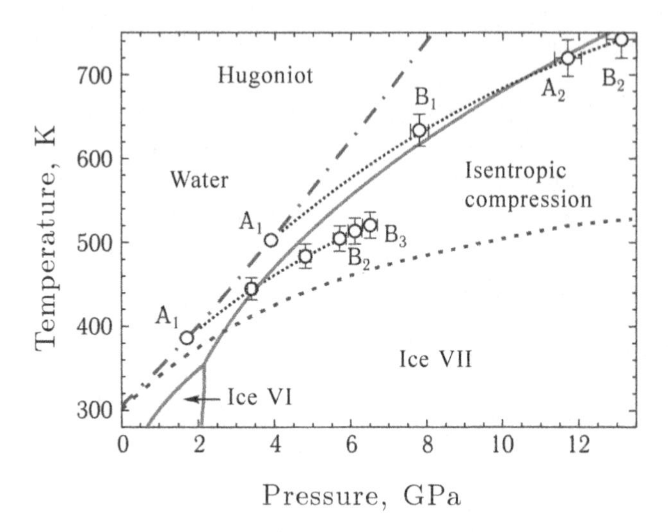

Fig. 7.15. Phase diagram of water in the temperature range above 0°C. Round markers connected by a dashed line show the change in the state of water under stepwise compression in experiments 2 (bottom row of dots) and 3. The meanings of the symbols A_i, B_i are explained in Fig. 7.18.

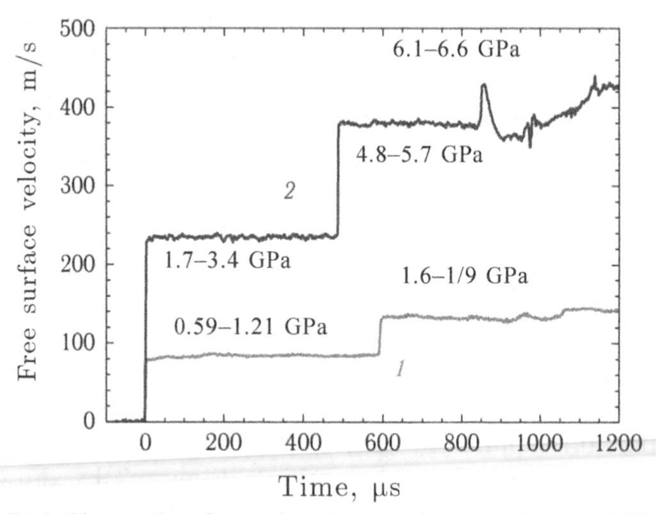

Fig. 7.16. The results of experiments at an impact velocity of 295 m/s and 1095 m/s. The base plates made of aluminium and copper, respectively, were used. The numbers indicate pressure before and after the reflection of the corresponding shock wave from the LiF window.

introduced into the sample, and the lid was a lithium fluoride monocrystal window. The shock wave in the base plate was created by impact of an aluminium plate, the speed of which varied from shot to shot.

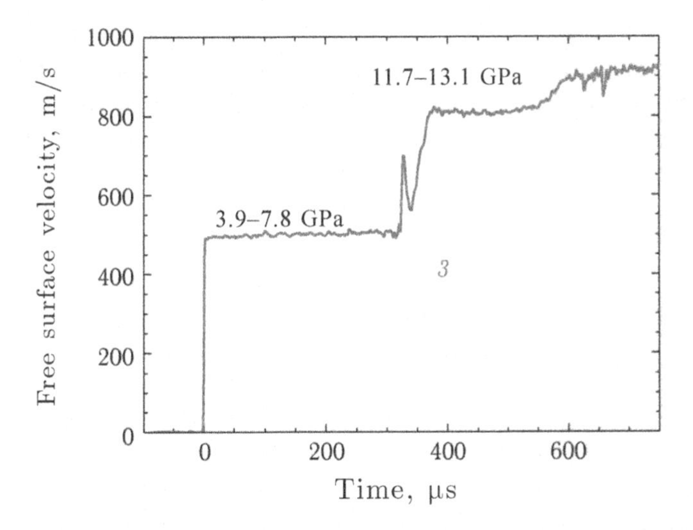

Fig. 7.17. The result of the experiment with the impacrt speed of 2600 m/s using a molybdenum base plate.

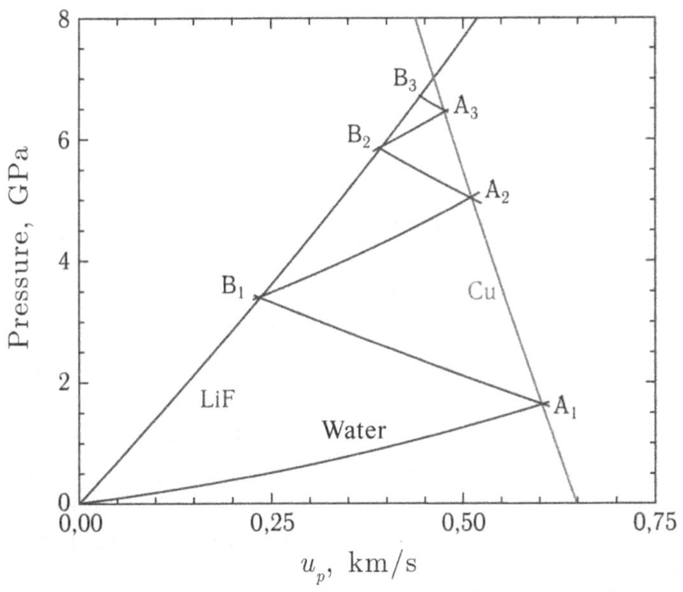

Fig. 7.18. The diagram of stepwise shock compression of water in experiment 2, shown in Fig. 7.16.

Figures 7.16 and 7.17 show the results of measurements in the form of velocity profiles of the contact surface of water and a lithium fluoride window as a function of time $u(t)$. Figure 7.18 explains the stepwise pressure build-up in water during multiple reflections of shock waves between the base plate and the window. The values of the particle velocity behind the shock wave were recorded on the contact surface of the LiF window with water – B_i points, while the parameters of shock waves approaching this surface correspond to the A_i points on the Hugoniot of braking the copper base plate.

In experiment 1 the pressure of 2.6 GPa is not reached, which corresponds to the intersection of the isentrop with the water – ice VII equilibrium line, water should not solidify in this experiment, the wave profile has no features. The rise time of the shock waves is at the limit of the temporal resolution of measurements, which indicates that the viscosity of the water is kept low during the stepwise compression process. In experiment 2, the first two steps have no features, and the exit to the window surface of the next shock wave is accompanied by a significant pressure relaxation. After relaxation, a gradual recovery of pressure is observed. Figure 7.15 shows the estimated state change during the stepwise compression process. It can be seen that the state on the second 'step' with a pressure of B_2 = 5.7 GPa is on the phase diagram far in the ice VII region, however, the transformation begins only when the pressure increases to 6.1 or 6.5 GPa (the uncertainty is related to the fact that it is unclear whether the transformation begins, at the base plate or at the surface of the window). The temperature deviation from the line of interphase equilibrium, that is, the supercooling of water before the start of transformation, is approximately 40 K. In experiment 3 with a pressure in the first shock wave in water of 3.9 GPa, the water –ice VII interphase boundary passes to the second step. Although at the same time the state of the substance is much less away from the interphase boundary than in experiment 2, pressure relaxation on the wave profile is clearly recorded and its fast recovery in the second step. New is the fact that the next wave of compression that has come to the surface of the window (the third step) is very blurred. A relatively long rise time of it can have two explanations: either we are dealing with a compression wave propagating over ice VII, having a viscosity much larger than water, or in this wave water turns into ice VII in about 40–50 ns. The second assumption is supported by the fact that the pressure of 13.1 GPa at the second recorded step corresponds exactly to the graphical analysis similar to those shown in Fig. 7.18, using the Hugoniot of water, that is, without taking into account its transformation into ice.

7.6. Evaporation when unloading from shock-compressed state

Usually, isentropes of unloading from a shock-compressed state are built according to a series of experiments in which the parameters of shock waves in low-impedance reference materials are located along the shock wave path following the substance under study. The specific internal energy and specific volume are then calculated from the measured Riemann isentrope $p(u)$. For condensed substances, the most important and interesting part of isentropes found in this way is the area of evaporation. The following describes a more visual and economical method of constructing isentropes in the field of evaporation, based on the recording of the acceleration of the foil of high-impedance material glued to the back of the plate of the test substance.

Figure 7.19 shows examples of such experiments with polymethyl methacrylate (plexiglass) as a test substance at shock compression pressures up to 34.5 GPa. In accordance with the calculations for the specified materials and foil thickness, its final velocity should be achieved after a ~200 ns process of multiple reflections of waves in it between the free surface and the surface of contact with the sample. This settling time is recorded in experiments with a pressure

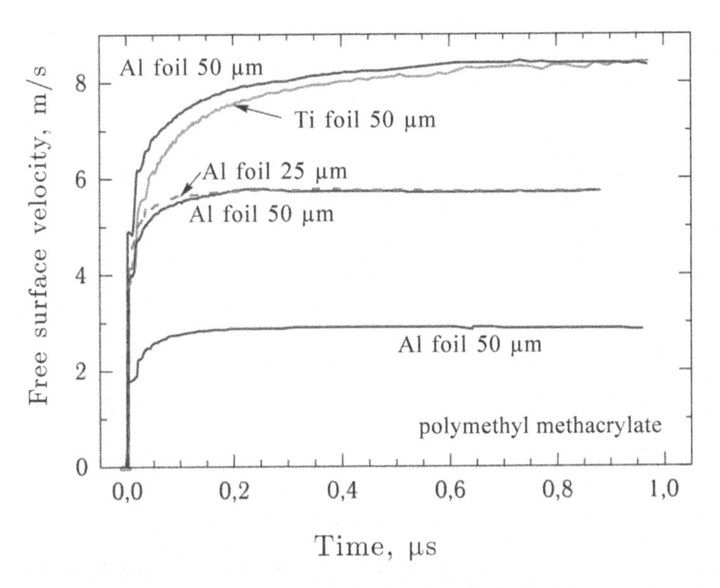

Fig. 7.19. The results of experiments with the recording of the acceleration of metal foil by shock-compressed polymethyl methacrylate (plexiglass).

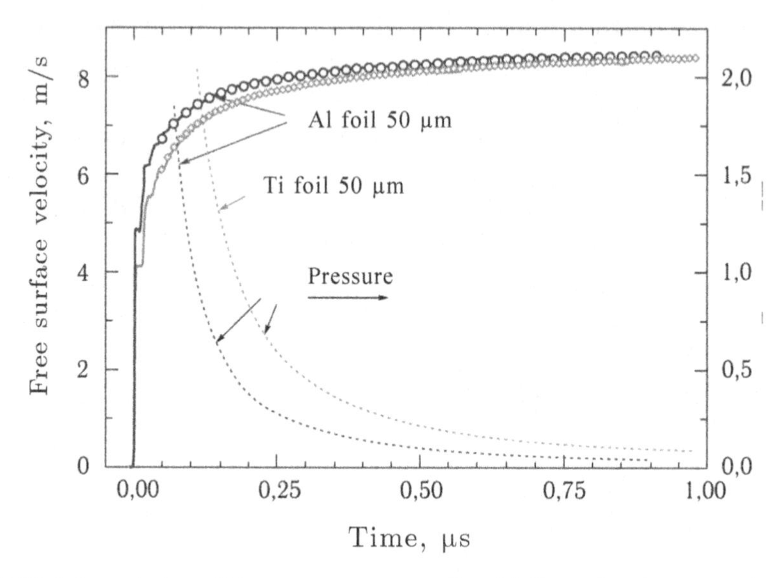

Fig. 7.20. Interpretation of the measurement results of foil acceleration at maximum shock compression pressure.

less than the maximum. However, the experiment with the maximum shock compression pressure demonstrates a much longer time to reach the maximum foil speed. This is a result of the evaporation of polymethyl methacrylate heated by a shock wave.

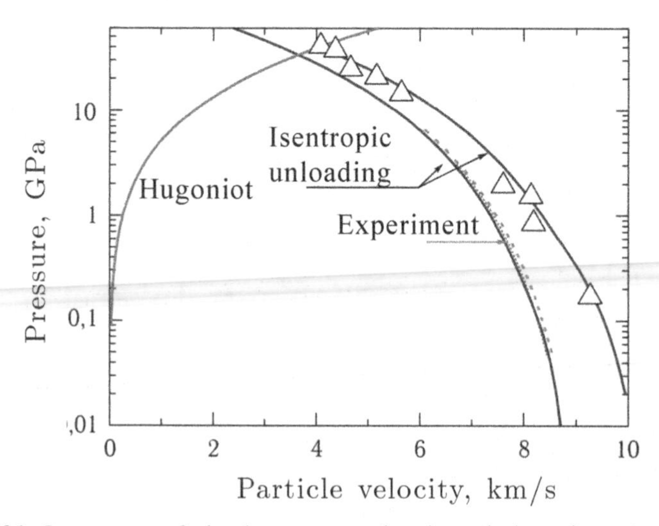

Fig. 7.21. Isentropes of shock-compressed polymethyl methacrylate unloading according to measurement results by the method of low-impedance reference barriers and the method of accelerated foil. Solid lines – calculation by the equation of state.

Interpretation of the measurement results is illustrated in Fig. 7.20. The upper part of the velocity history is approximated by a smooth function $u_f(t)$ with monotonously decreasing first and second derivatives. Then, knowing the mass per unit area of the foil, the pressure of the evaporated substance is determined in accordance with Newton's law $p(t) = \rho\delta du_f(t)/dt$, where ρ and δ are the density and thickness of the foil. The values of pressure p and particle velocity $u = u_f$, taken at the same moment of time t, correspond to a point on the isentrope of unloading shock-compressed polymethyl methacrylate.

The results of the analysis are shown by dashed and dotted lines in the p–u diagram in Fig. 7.21. It can be seen that experiments with the aluminium and titanium foils with different masses are in good agreement with each other and with the plexiglas state equation, which also takes into account the data of experiments with low-impedance reference barriers.

8 Determination of macrokinetic regularities of the transformation of explosives in shock and detonation waves

To calculate detonation processes, information is needed on the equations of state of explosives and explosion products and on the macrokinetic relationships for transformation of explosive substances in explosion products in shock and detonation waves. This information is currently obtained by means of continuum mechanics from the analysis of the evolution of shock compression pulses in the high explosive (HE).

The establishment of steady detonation is preceded by the process of its initiation. Although the initiating impulse may have a different nature, the formation of a detonation wave occurs through the formation of a shock wave. As a result of a chemical reaction in shock-compressed explosives, its intensity increases as it propagates, and the shock wave transforms into a detonation wave. In homogeneous explosives initiation of detonation by a shock wave occurs according to the mechanism of adiabatic thermal explosion. In the shock wave, the temperature of explosives rises, which excites the exothermic chemical reaction. Under adiabatic conditions, the heat released is spent on the heating of the reactant, which accelerates the reaction. Thus, the rate of the process increases exponentially – a thermal explosion occurs. As a result, after the induction time, the pressure increases sharply, a compression wave is formed in the compressed substance, which then overtakes the front of the initiating shock wave and then propagates along the uncompressed substance in the form of a steady detonation wave.

The mechanism of a homogeneous thermal explosion reasonably agrees with the experimental data for liquid explosives, but does not explain the patterns of initiation of detonation of solid explosives. For example, at a pressure in the shock wave of 5.5 GPa, the time of decomposition of desentized RDX is of the order of a microsecond, although its temperature is only about 110°C. This is a hundred degrees below the flash temperature of RDX with a 5 min exposure under normal conditions.

The fact is that the solid explosives, as a rule, are not homogeneous and contain various defects – pores, cracks, inclusions, grain boundaries and crystal microdefects in the structure. Physical inhomogeneity leads to an uneven distribution of energy during the rapid compression of an explosive by a shock wave. Part of the energy of the shock wave is localized in individual 'hot spots', which leads to the excitation of an exothermic decomposition reaction in them. The heat released during the development of a chemical reaction is partially transmitted to the surrounding substance and partially remains in the nucleus, increases its temperature and thereby accelerates the process of decomposition. If heat generation prevails, then a progressive rise in the temperature of the nucleus and reaction rate occurs – thermal self-ignition occurs. The less heated mass of the explosives is ignited by the hot spot and then burns in the combustion waves propagating from the nucleus. The coalescence of the hot spots is followed by the beginning of the the degressive stage of burning.

Since the heat losses are proportional to the surface area of the nucleus, and the amount of energy released is proportional to its temperature and volume, it is clear that flashes are possible in those nuclei that are larger than a certain limit for a given temperature. The lower the temperature of the region, the larger must be its size to ensure a sustainable ignition of explosives. With an increase in the amplitude of the shock wave, the temperature of the hot spots increases. Consequently, a larger number of nuclei that can ignite the surrounding explosives are formed behind the front of a stronger shock wave.

The largest voids in the starting material are already eliminated at relatively low pressure shock pressures; however, only part of them forms the reaction nuclei. Homogenization of the explosives in weak shock waves reduces its sensitivity to subsequent stronger shock wave effects. Experiments also show that the initiation of the reaction is suppressed by the blurring of the compression wave. The dependences of the duration of the chemical spike of the detonation wave and the failure diameter of detonation on the structure of explosives indicate that the process of energy release has a heterogeneous, focal nature not only in initiating shock waves, but also in the detonation conditions.

The hypothesis of the initiation of an explosion in the hot spots was advanced and justified by F.P. Bowden and A.D. Ioffe in studies of the excitation of an explosion of condensed explosives by mechanical impact. The effects of microcumulation on hard impurity particles and in the pores, plastic work on the periphery of the pores, friction between particles, cracking of explosive grains during compaction, adiabatic compression of gas inclusions and others are discussed as possible mechanisms for the formation of hot spots. The greatest progress in the theoretical analysis of the mechanisms of the formation of 'hot spots' has been achieved in the calculations of the viscous–plastic heating of a substance in the vicinity of a collapsing spherical pore. The size of the hot spots in the shock-compressed substance is proportional to the size of the original inhomogeneities. Therefore, the coarse-grained explosives have a lower initiation threshold than the finely dispersed ones. However, since the concentration of the regions is higher in the fine explosives, after initiating the reaction the transition to detonation occurs more quickly in it. The lower limit of the possible values of the pressure of the initiating shock wave is, apparently, the yield strength of this explosive.

The presence of a spectrum of potential nuclei qualitatively changes the dynamics of shock-wave initiation of detonation of inhomogeneous explosives. In solid explosives, a pronounced induction period is not observed, the initiating shock wave transforms into detonation motion relatively smoothly, without jumps and over compression. An example of the recording [75] of the transformation of a shock wave into detonation is shown in Fig. 8.1. The development of the process of explosive transformation leads to an increase in the pressure behind the jump and a continuous amplification of the shock wave until it reaches the detonation mode. The amplification of the shock wave can occur even in the absence of a pressure rise region behind the shock wave. Currently, there are two approaches to solving the problem of obtaining macrokinetic information.

In the first case, information on the formal kinetics of the transformation of the initial substance into the explosion products

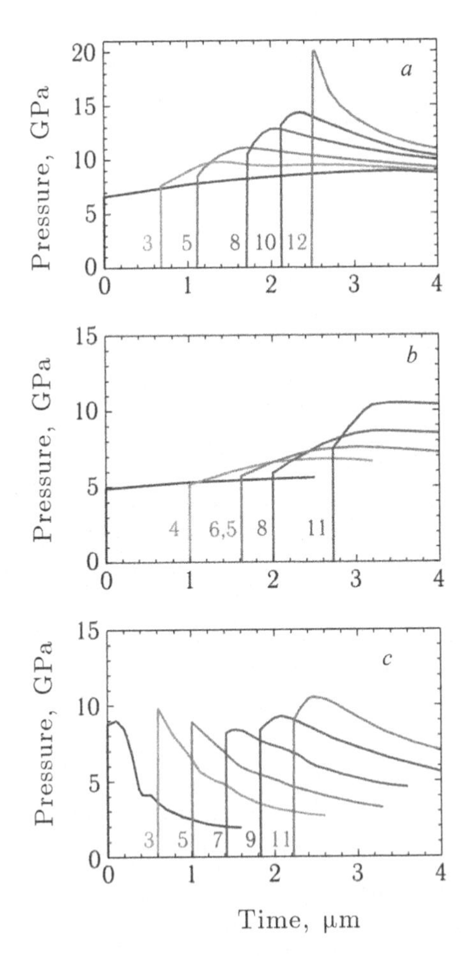

Fig. 8.1. Evolution of shock compression pulses of various amplitudes and durations in cast TNT. Measurements with manganin pressure gauges.

is extracted directly from an analysis of the evolution of a one-dimensional shock compression pulse. The second approach is to use experimental data to verify the mathematical model of the process and adjust its parameters.

A direct way to obtain kinetic data is to interpret, based on the laws of conservation fields of pressure or particle velocities, constructed from the measurement results of the corresponding wave profiles in several sections of the explosive sample. As an example Fig. 8.2 shows the pressure field for the experimental series shown in Fig. 8.1 *a*.

Using the conservation laws, one can go from the pressure field to the particle velocity field and from the velocity field to the compression field. Several analysis methods have been proposed that minimize the processing error for such a recalculation. From these

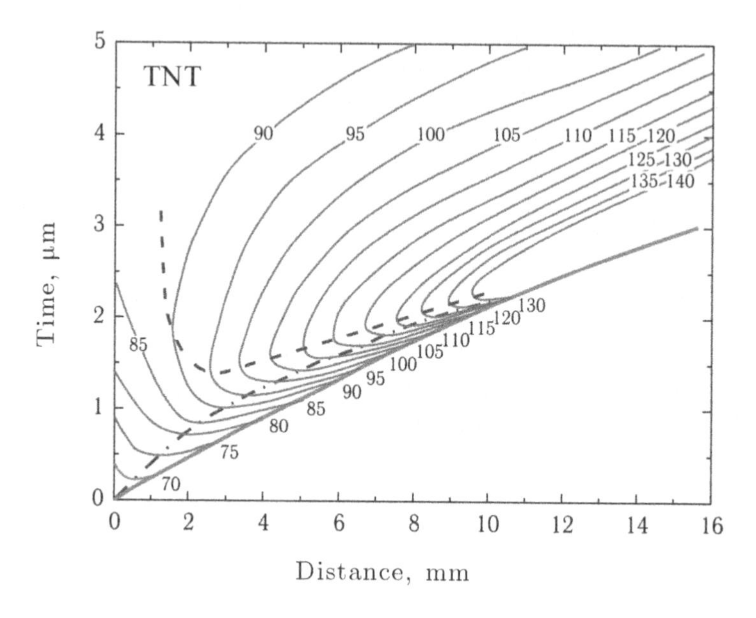

Fig. 8.2. The pressure field for the experimental series shown in Fig. 8.1 *a*. The numbers indicate the pressure values at the corresponding isobars.

data, the course of the change in the state of the selected particles of a substance in the pressure–specific volume coordinates is restored. Each point of such a trajectory corresponds to a certain point in time. Assuming the additivity of the specific volumes and energies of the original explosive and its decomposition products, each point of the trajectory of the state change corresponds to a certain concentration of explosion products α. Thus, for each selected particle, the law of change in the depth of decomposition with time is determined. An example of this approach is illustrated in Figs. 8.3 and 8.4.

Figure 8.3 shows the trajectories of the change of state behind a shock jump in cast TNT, obtained as a result of processing a series of experimental pressure profiles, which are shown in Fig. 8.1. As the decomposition of explosives develops, the states of the particles deviate from the Hugoniot of the unreacted HE and approach the isentrope of the explosion products. To illustrate the speed of the process, time marks with an interval of 0.5 μs are plotted on the state transition paths.

Figure 8.4 shows the concentration curves describing the change in the depth of decomposition α of selected particles with time after the passage of the shock wave front. The dependences $\alpha(t)$ are plotted under the assumption of additivity of the mixture components.

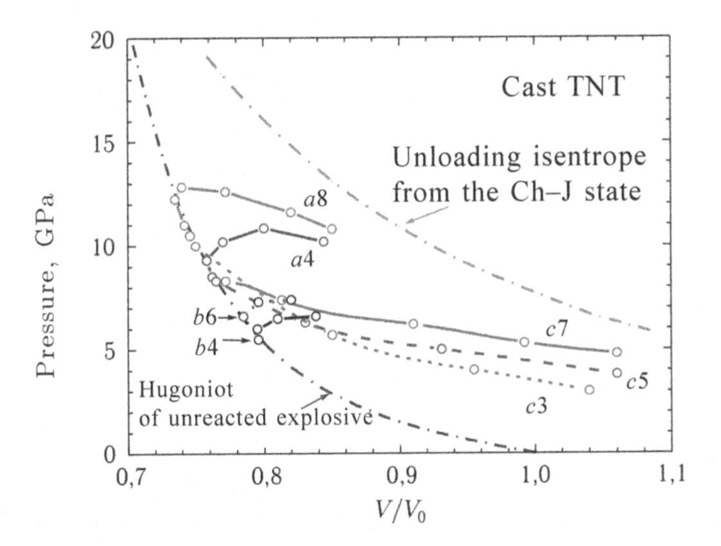

Fig. 8.3. The trajectories of changes in the state of the selected particles, constructed according to Fig. 8.1. Markers are placed at regular intervals of 0.5 μs.

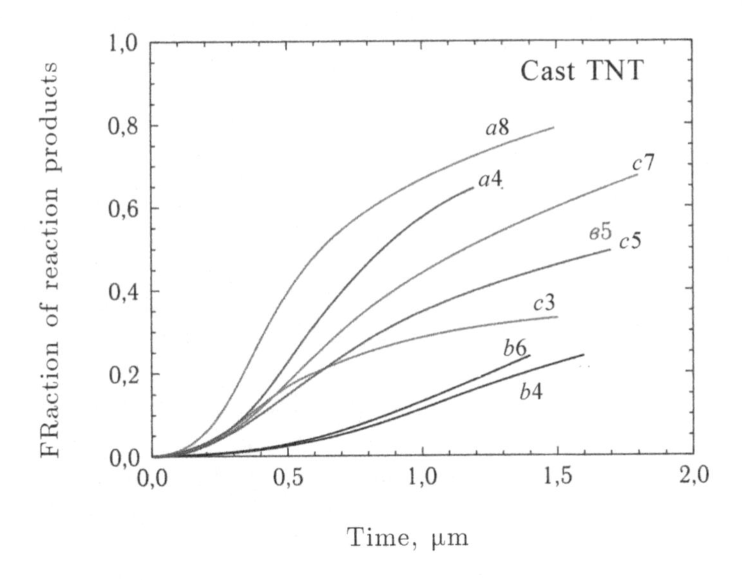

Fig. 8.4. Concentration curves plotted according to Fig. 8.3. The experimental series is shown in Fig. 8.1 and the distance travelled by the shock wave.

Although the error in determining the depth of decomposition is significant, a set of kinetic curves makes it possible to judge the general laws of the process. It is seen, in particular, that immediately behind the pressure jump the decomposition rate is close to zero with increasing decomposition depth, it grows and, in the case when the pressure change is not too large, passes through a maximum in the vicinity of $\alpha \sim 0.2$–0.3. The shift of the maximum decomposition rate to the beginning of the process is easy to understand, assuming that the decomposition nuclei are concentrated on the surface of explosive grains and their number is such that the distance between them is smaller than the grain size. A sharp decrease in pressure, as can be seen from curves *c*3 and *c*5, leads to a drop in the rate of decomposition.

The described approach is associated with the need to carry out a large volume of time-consuming experiments with increased demands on the accuracy of measurements. A more common way to obtain macrokinetic information is based on a combination of measurements with mathematical modelling of the experimental situation. With this approach, the central issue is the choice of a rational kinetic model of decomposition of heterogeneous explosives. Unfortunately, the lack of information about the properties of substances, size, shape and mechanism of the formation of hot spots makes it difficult or makes it too bulky to describe in detail and consistently the excitation and propagation of the reaction within a single and acceptable model. The absence of a rigorous, physically sound model of the emergence and development of 'hot spots' is partially compensated by a variety of semi-empirical models, based on the most general ideas about the nature of the process. The coefficients of relations describing the dependence of the rate of decomposition of explosives (that is, macrokinetics equations) on the main parameters of the state are fully or partially subject to experimental determination.

The macrokinetic relation discussed below is based on the following model representations. The decomposition of shock-compressed solid explosives originates in 'hot spots', which are formed as a result of the energy concentration of the shock wave in the vicinity of the initially existing micropores, grain boundaries, inclusions and other inhomogeneities. After nucleation, the reaction propagates from the activated hot spots to the surrounding matter in the form of combustion waves. It is essential that the hot spots form in the shock wave, so that the parameters of the shock wave determine the number of reaction sites and the corresponding average decomposition rate. Thus, the macroscopic rate of transformation of an explosive into explosion products is proportional to the density of activated nuclei (which depends on the intensity of the shock wave

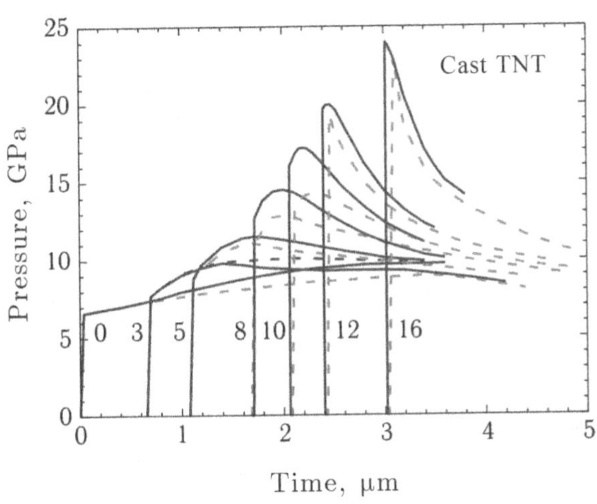

Fig. 8.5. The results of computer simulation (dashed lines) of the experimental series shown in Fig. 8.1 *a*.

and the initial structure of explosives), the velocity of the combustion waves propagating from the nuclei to the surrounding matter (which is proportional to pressure), and the effective area of combustion, which changes as burnout explosives. Since information on the properties of explosives is usually very limited and insufficient for development of a complete closed model, the macrokinetic constitutive relation for practical applications should be simple and contain the minimum number of parameters related to the properties of a particular explosive.

In the examples of computer simulation shown below, a macrokinetic ratio was used in the form [76]

$$\partial\alpha/\partial t = k \cdot (E_{sh} - E_{thre}) \cdot p \cdot \alpha^{\gamma} \cdot (1 - \alpha)^{1-\gamma}, \tag{8.1}$$

where α is the mass fraction of the decomposition products of explosives, E_{sh} is the increment of the specific internal energy in the shock wave passing through the explosive particle under study, k and γ are material constants, E_{thre} is the threshold specific internal energy of the initiating shock wave. It is assumed that decomposition reactions in shock-compressed explosives are not initiated if $E_{sh} < E_{thre}$. A certain disadvantage of the macrokinetic relation in the form of (8.1) is the need to trace the shock discontinuity in the calculations and to accurately determine its parameters. An example of the results of such calculations in comparison with experimental data for cast TNT is shown in Fig. 8.5.

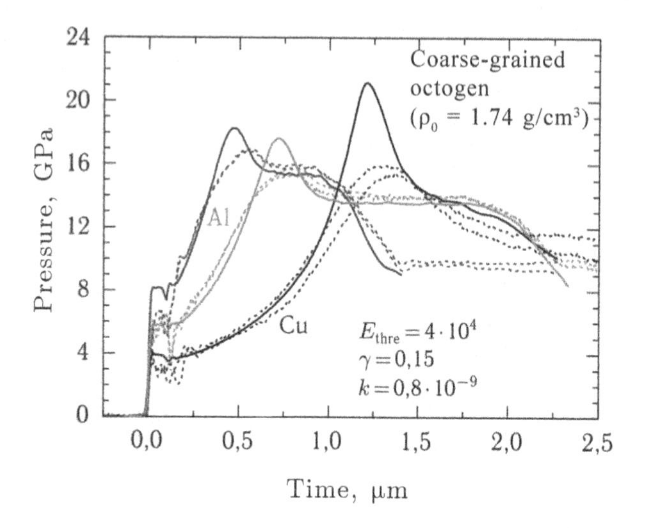

Fig. 8.6. Pressure histories at the interface between the inert base plate and the HMX sample, measured in experiments and obtained as a result of numerical simulation of experiments. The material of the intermediate base plate (Al, Cu) between the sample and the aluminium impactor is indicated. The impact velocity is 1.17 km/s or 1.55 km/s. Measurements were taken with manganin gauges placed in a teflon insulation film (taken into account in the calculations).

In Fig. 8.6, the results of computer simulation of shock-wave detonation initiation based on this macrokinetic ratio are compared with experimental data for HMX. In the experiments, pressure histories $p(t)$ were measured at the interface of the explosive under investigation with an inert base plate through which a shock compression impulse generated in the base plate by the impact of a plate was introduced into the sample. The variation of the impact velocity and the base plate material varied the initial pressure in the explosive sample. The pressed HMX was investigated. The formulation of experiments is described in [77]. There is a reasonable agreement between the results of calculations and experimental data on the initial and final sections of the wave profiles. The overall shape of the wave profiles is also consistent. There is a difference in the maximum pressure values, which can be explained by the violation of the one-dimensionality of the flow during the rapid acceleration of the shock wave in the explosive.

When using the macrokinetic relation (8.1), it is necessary to track the propagation of a shock wave in explosives. In Fig. 8.7, one of the calculated pressure histories at the interface between the base plate and the sample of explosives (experiment with a copper

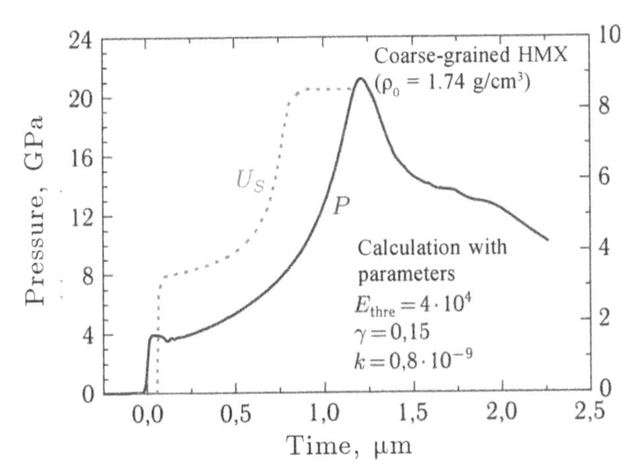

Fig. 8.7. Comparison of the calculated pressure evolution at the interface between the HMX sample and the copper base plate and the velocity of the shock wave in the HMX. The constant velocity of the shock wave in the final section is the HMX detonation velocity.

base plate in Fig. 8.6) and the velocity of the shock wave in the explosive as a function of time are compared. It can be seen that in the near-boundary layers the energy release and, accordingly, the pressure increase lasts longer and ends only after the shock wave has passed into the detonation one.

Figure 8.8 compares the results of measurements and computer simulation of the detonation wave. In the experiments, aluminium foil was pressed onto the surface of the HMX sample, behind which a water 'window' was located. The velocity history of the contact surface of the foil reflector for the laser radiation of a VISAR instrument with a water window was recorded. Here a good agreement of the calculations with the experimental data is also obtained. Thus, by relatively simple experiments, the entire practically important pressure range is covered – from pressures close to the initiation threshold to detonation pressure. As a result, information on the macrokinetic patterns of a chemical reaction was obtained by means of mechanics, which under normal conditions are investigated by much more complex physical and chemical methods. Studies of macrokinetics of decomposition of explosives in shock waves, started in the mid-seventies, attracted much attention and aroused heated discussions, but have now been significantly reduced due to the limited number of explosives for which they are justified.

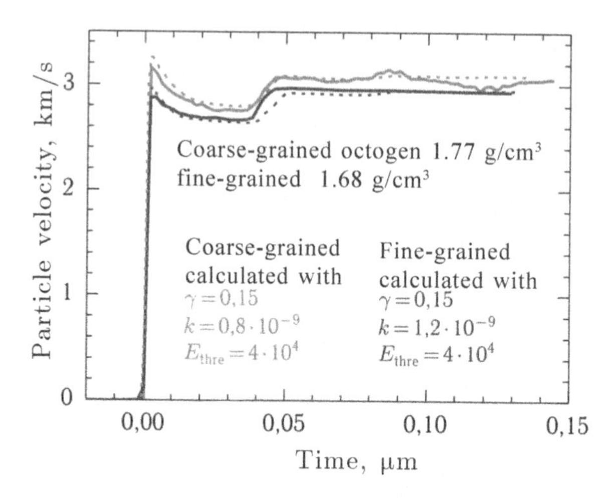

Fig. 8.8. Results of modelling (dashed lines) of experiments on the measurement of the zone of a chemical reaction of a steadty detonation wave in a pressed HMX with different grain sizes. The velocity histories of the surface of the foil reflector with a thickness of 200 µm in contact with a water window are shown.

According to the theory of detonation of Ya.B. Zel'dovich, the detonation wave should contain a peak of pressure in the frontal part – 'chemical peak', in which the transformation of an explosive into explosion products occurs. The chemical peak was repeatedly recorded experimentally with the detonation of a variety of explosives. Later, however, data appeared [86, 87] that for some explosives the detonation wave does not contain chemical peak or even instead of the chemical peak it shows a slight increase in pressure. The detonation wave without the chemical peak takes place at a high explosive density, and at lower densities, the detonation wave contains a chemical peak, whose amplitude gradually decreases with increasing density. The disappearance of the chemical peak is explained by the departure of the reaction into a shock wave, that is, it is assumed that all conversion of explosives into explosion products or its main part occurs directly in the process of shock compression. This explanation, however, was not supported by any calculations in the framework of a consistent model or independent measurements. On the other hand, the separation of a detonation wave into a shock wave in which the compression of a substance occurs so quickly that a noticeable transformation does not occur in time, and the reaction zone following it turned out to be fruitful. As was shown above, this separation allows us to describe the

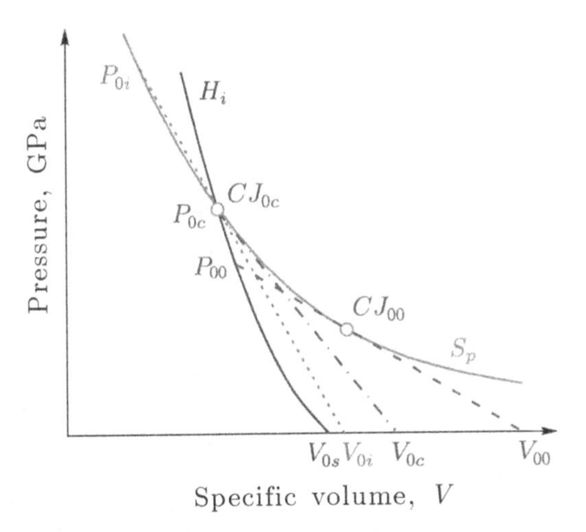

Fig. 8.9. State diagrams in detonation waves in the case of intersecting adiabats of the original explosive and explosion products. For clarity, all the states of the products of explosion at different incident densities of HE are represented by a single $p(V)$ curve S_p, and the states of the original explosive are represented by the H_i curve.

development of detonation in the entire pressure range using unified equations of state and a single macrokinetic relationship.

So far, we have assumed that the adiabat of gaseous explosion products is entirely located to the right of the Hugoniot of unreacted explosives. Let us now admit the possibility of intersection of the adiabats of the original explosive and the explosion products. This assumption in fact does not contradict anything, especially if we take into account the possibility of the formation of a dense carbon–diamond phase in the explosion products. The situation with the intersection of adiabats for explosives of different densities is illustrated schematically in Fig. 8.9, where for clarity, all the states of the explosion products are represented by a single $p(V)$ curve S_p, and the states of the original explosive are represented by the H_i curve. At a low density of explosives corresponding to a specific volume V_{00}, detonation with a chemical peak occurs, since the intersection of adiabats occurs at higher pressures than pressures at the wave front and in the Chapman–Jouguet plane for this density. With increasing density, the relative amplitude of the chemical peak (the pressure difference between the state behind the shock wave at point P_{00} and the state in the Chapman–Jouguet plane, described by point CJ_{00}) should decrease, as was observed in experiments. At a certain threshold value of the initial specific volume V_{0c}, the tangency point falls at the intersection point of the Hugoniots

Let us consider in more detail what happens with a further increase in the density of explosives. The initiating shock wave grows until the pressure behind its front reaches the value P_{0c} corresponding to the intersection point of the adiabats of the explosive and the explosion products. As this point is approached, the difference between the explosive volumes and explosion products $V_p - V_i$ decreases to zero, as a result of which the effect of energy release on the evolution of the shock wave stops. With the passage through the intersection point, the fulfillment of the Chapman–Jouguet condition becomes impossible, since the Rayleigh line describing the states in the shock wave in unreacted explosives cannot pass below the Hugoniot. Detonation with pressure at the front above the P_{0c} point cannot be self-sustaining, since all isentropes of the mixtures cross the wave beam, the flow is subsonic. As a result, a rarefaction wave from the explosion products can penetrate to the shock wave and cause its attenuation up to the point of intersection of the adiabats. Thus, the adiabat intersection in this case determines the detonation parameters instead of the Chapman–Jouget selection rule.

Conclusion

The presented results of new experiments confirm the informativeness and fruitfulness of shock-wave studies of the strength properties, polymorphism and metastable states of materials and substances with extremely short durations of mechanical action. In recent years, the temporary range of studies, thanks to which it became possible to carry out measurements actually at the limit of applicability of continuum mechanics and enter the range of spatial and temporal scales that previously had to be dealt with only when modelling the movement of a substance using molecular dynamics methods have been been greatly expanded.

The obtained new systematic data and temperature–velocity dependences of the resistance of high-speed deformed structures with different crystal structures form the basis for developing comprehensive wide-range determinant relations necessary for the calculation of intense pulsed effects on materials and structures. The range of materials in which an anomalous increase in the stress of high-speed plastic flow at elevated temperatures is possible is determined. New information has been obtained about the laws of reproduction of elementary plastic deformation carriers – dislocations in the early stages of deformation. It was found that at high strain rates, hardening inclusions in alloys can be sources of dislocation and have a softening effect. It is shown that the partial melting of the alloy is not necessarily accompanied by the immediate loss of its strength properties. Deep penetration into the region of negative pressures opens up the possibility of studying phase transitions and polymorphic transformations under tension, and, possibly, allows us to speak about the formation of the physics of negative pressures. The observed fracture waves in shock-compressed glasses are a type of regular failure, the study of which can clarify the mechanisms of earthquakes and collapses of mine workings.

A number of specific features of the dynamics of elastoplastic shock compression waves in relaxing media have been discovered, which are still awaiting reflection and theoretical description. Unfortunately, the absence of any complete theory of the phenomenon limits the amount of information obtained from the analysis of the dynamics of elastoplastic shock compression waves. The author hopes that the presented review will stimulate theoretical work in this direction.

The recognized largest achievements of modern physics of shock waves in solids were obtained using unique and expensive experimental facilities, such as the NIF laser complex at the Livermore National Laboratory, a pulsed power plant for generating Z-pinch in the Sandia Laboratory, etc. The experimental results briefly discussed in this the book, on the other hand, were obtained using more or less standard and relatively inexpensive equipment available for ordinary academic institutions and university laboratories.

An abnormal increase in the plastic flow stress with temperature at high strain rates might have been expected, but was not predicted; the effect was detected more or less by chance. Failure waves are another example of an unexpected find. They not only explained the existing contradiction in the data on the spall strength of glass obtained by different authors, but also discovered a way to implement research of metastable states of glass. These new phenomena and effects were discovered when we simply tried to expand the range of experimental parameters and find explanations for some of the observed details of the processes accompanying the shock compression of solids. All this means that many more unusual and interesting things await us on the path of research of shock-wave phenomena in solids, and any researcher may be lucky enough to discover them.

References

1. Zel'dovich Ya.B., Raiser Yu.P., Physics of shock waves and high-temperature hydrodynamic phenomena. Moscow, Nauka, 1966.
2. Baum F.A., Orlenko A.P., Stanyukovich K.P., Chelyshev V.P., Shekhter B.I., Physics of the explosion. Moscow, Nauka, 1975.
3. Zhernokletov M.V. (ed.), Methods for studying the properties of materials under intense dynamic loads. - Sarov: Federal State Unitary Enterprise RFNC-VNIIEF, 2003.
4. Kanel G.I., Razorenov S.V., Utkin A.V., Fortov V.E., Shock-wave phenomena in condensed media. - Moscow, Janus-K, 1996.
5. Ivanov A.G. and Novikov S.A., Pribory i tekhnika eksperimenta. 1963. Vol. 7, No. 1. P. 135–138.
6. Barker L.M. and Hollenbach R.E., J. Appl. Phys. 1972. Vol. 43 (11). P. 4669–4675.
7. Fowles R. and Williams R. F., J. Appl. Phys. 1970. Vol. 41 (1). P. 360–363.
8. Seaman L., J. Appl. Phys. 1974. Vol. 45 (10). P. 4303–4314.
9. Duvall G.E., Propagation of the stress-relaxing medium, In: Stress Waves in Anelastic Solids, edited by H. Kolsky and W. Prager. - Berlin: Springer-Verlag, 1964. - P. 20.
10. Swegle J.W. and Grady D.E., J. Appl. Phys. 1985. Vol. 58. P. 692-701.
11. Novikov S.A., Divnov I.I., Ivanov A.G., Fiz. Met. Metalloved. 1964. Vol. 25, No. 4. P. 608–615.
12. Stepanov G.V., Probl. prochn. 1976. No 8. P. 66–70.
13. Romanchenko V.I. and Stepanov G.V., PMTF. 1980. No 4. P. 141–147.
14. Antoun T., Seaman L., Curran D. R., Kanel G. I., Razorenov S.V., Utkin A.V. Spall Fracture. - New York: Springer, 2003.
15. Garkushin G.V., Kanel G.I., Razorenov S.V., Fiz. Tverd. Tela. 2010. V. 52, No. 11. pp. 2216–2222.
16. Zaretsky E.B. and Kanel G.I., J. Appl. Phys. 2012. Vol. 112. P. 073504.
17. Kanel G.I., Razorenov S.V., Baumung K., and Singer J., J. Appl. Phys. 2001. Vol. 90 (1). P. 136–143.
18. Savinykh A.S., Kanel G.I., Razorenov S.V., Pis'ma v ZhTF. 2011. V. 37, No. 7. P. 8–15.
19. Kanel G. I., Bogach A.A., Razorenov S.V., Savinykh A.S., Chen Z. and Rajendran A., Wave Study of the Failure Wave Phenomenon in Brittle Materials, In: Shock Compression of Condensed Matter - 2003, Eds. M.D. Furnish et al., AIP CP. 2004. Vol. 706. P. 739–742.
20. Savinykh A.S., Kanel G.I., Razorenov S.V., ZhTF. 2010. V. 80, No. 6. P. 85–89.
21. Parkhomenko I.P., Utkin A.V., Spall strength of plexiglass, in: Studies of the proper-

ties of a substance in extreme conditions. Ed. V. Ye. Ortov et al. - Moscow, IVTAN, 1990. P. 126–130.
22. Kalmykov Yu.B., Kanel G.I., Parkhomenko I.P., Utkin A.V., Fortov V.E., PMTF. 1990. No 1. P. 126–130.
23. Utkin A.V., Sosikov V.A., Bogach A.A., Fortov V. E. Tension of liquids by shock waves, In: Furnish M.D., Gupta Y.M., and Forbes J.W. (eds.) Shock Compression of Condensed Matter - 2003, AIP Conference Processions. Melville, New York. 2004. Vol. 706. p. 765-770.
24. Kanel G.I., Savinykh A.S., Garkushin G.V., Razorenov S.V., Pis'ma v ZhETF. 2015. V. 102, no. 8. pp. 615–619.
25. Zaretsky E.B., J. Appl. Phys. 2016. Vol. 120. P. 025902.
26. Ashitkov S.I., Komarov P.S., Ovchinnikov A.V., Struleva E.V., Agranat M.B., Pis'ma v ZhETF. 2016. V. 103, No. 8. P. 611–616.
27. Fayzullin M.Z. and Skripov V.P., TVT. 2007. V. 45, No 6. P. 881–884.
28. Utkin A.V., PMTF. 1997. V. 38, No 6. P. 157–167.
29. Sin'ko G.V. and Smirnov N.A., Pis'ma v ZhETF. 2002. V. 75, No. 4. P. 217–219.
30. Cerny M. and Pokluda J., Phys. Rev. B. 2007. Vol. 76. P. 024115.
31. Joshi K.D. and Gupta S.C., High Pressure Research. 2007. Vol. 27 (2). P. 259–268.
32. Friakyz M., Sob M. and Vitek V., Phil. Mag. 2003. Vol. 83. P. 3529-3537.
33. Sinko G.V. and Smirnov N.A., Letters to JETP. 2004. V. 79 (11). Pp. 665–669.
34. Speedy R.J., J. Phys. Chem. 1982. Vol. 86. P. 982–991.
35. Netz P.A., Starr F.W., Stanley H.E., Barbosa M.C., J. Chem. Phys. 2001. Vol. 115. P. 344–348.
36. 36. Ogorodnikov V.A., Borovkova E.Yu., Erunov S.V., Fiz. Gor. Vzryva. 2004. V. 40, No. 5. 109–117.
37. Kanel G.I., Razorenov S.V., Bogatch A.A., Utkin A.V., Fortov V. E., and Grady D. E., J. Appl. Phys. 1996. Vol. 79 (11). P. 8310–8317.
38. Bogach A.A., Kanel G.I., Razorenov S.V., Utkin A.V., Protasova S.G., Sursaeva V. G., Fiz. Tverd. Tela. 1998. V. 40 (10). Pp. 1849–1854.
39. Dash J.D., Review of Modern Physics. 1999. Vol. 71 (5). P. 1737–1743.
40. Besold G. and Mouritsen O.G., Phys. Rev. B. 1994. Vol. 50 (10). P. 6573–6576.
41. Zaretsky E.B. and Kanel G.I., J. Appl. Phys. 2012. Vol. 112. P. 053511.
42. Alshits V.I., Indenbom V.L., Usp. Fiz. Nauk. 1975. T. 115, No. 1. P. 3.
43. Asay J. R., Fowles G. R. and Gupta Y., J. Appl. Phys. 1972. Vol. 43. P. 744.
44. Chhabildas L.C. and Asay J.R., J. Appl. Phys. 1979. Vol. 50, No 4. P. 2749.
45. Sakino K., J. Phys. IV. France. 2000. Vol. 10. P. Pr9-57–62.
46. Kanel G.I., Razorenov S.V., Garkushin G.V., Pavlenko A.V., Malyugina S.N., Solid State Physics. 2016. V. 58, No. 6. pp. 1153–1160.
47. Kanel G.I., Savinykh A.S., Garkushin G.V., Razorenov S.V., Teplofizika Vys. Temperatur. 2017. Vol. 55, No. 3. P. 380–385.
48. Kruger L., Meyer L., Razorenov S.V., Kanel G.I., International Journal of Impact Engineering. 2003. Vol. 28 (8). P. 877–890.
49. Kanel G.I., Razorenov S.V. and Fortov V.E., Journal of Physics: Condensed Matter. 2004. Vol. 16, No. 14. P. S1007.
50. Ashitkov S.I., Komarov P., Agranat M.B., Kanel G.I., Fortov V.E., Pism'a v ZhETF. 2013. V. 98 (7). P. 439.
51. Zaretsky E.B. and Kanel G.I., Journal of Applied Physics. 2014. Vol. 115. P. 243502.
52. Garkushin G.V., Ignatova O.N., Kanel G.I., Meyer L., Razorenov S.V., Izv. RAN. Mekh. Tverd. Tela. 2010. No 4. P. 155–163.
53. Zaretsky E.B. and Kanel G.I., J. Appl. Phys. 2012. Vol. 112. P. 073504.

54. Zaretsky E.B. and Kanel G.I., J. Appl. Phys. 2015. Vol. 117. P. 195901.
55. Kanel G.I., Garkushin G.V., Savinykh A.S., Razorenov S.V., de Resseguier T., Proud W.G., and Tyutin M.R. J. Appl. Phys. 2014. Vol. 116. P. 143504.
56. Johnson J.N., J. Appl. Phys. 1972. Vol. 43 (5). P. 2074.
57. Zaretsky E.B. and Kanel G.I., J. Appl. Phys. 2013. Vol. 114. P. 083511.
58. Kanel G.I., Razorenov S.V., Garkushin G.V., Savinykh A.S., and Zaretsky E.B., J. Appl. Phys. 2015. Vol. 118 (4). P. 045901.
59. Griffith A.A. The theory of rupture. Proc. 1st int. Con. Appl. Mech. (Delft). 1924. P. 55–63.
60. Kanel G.I., Bogatch A.A., Razorenov S.V., Zhen Chen., J. Appl. Phys. 2002. Vol. 92 (9). P. 5045-5052.
61. Savinykh A.S., Kanel G.I., Cherepanov I.A., Razorenov S.V., Zh. Tekh. Fiziki. 2016. V. 86, No. 3. P. 70–76.
62. Kanel G.I., Bless S.J., Savinykh A.S., Razorenov S.V., Chen T., and Rajendran A., J. Appl. Phys. 2008. Vol. 104. P. 093509.
63. Paris V. E., Zaretsky E.B., Kanel G. I., and Savinykh A.S., Diagnostics of Ductility, Failure, and Compaction of Ceramics under Shock Compression. In: Shock Compression of Condensed Matter - 2003, Eds. M.D. Furnish et al. AIP CP. 2004. Vol. 706. P. 747–750.
64. Savinykh A.S., Kanel G.I., Razorenov S.V., Rajendran A., Compressive fracture of brittle materials under divergent loading loading. In: Shock Compression of Condensed Matter - 2005, Eds .: M.D. Furnish et al. - New York: American Institute of Physics, 2006. P. 888–891.
65. Paris V.E. and Zaretsky E.B., Study of Compressive Failure of Alumina in Impact Experiments with Divergent Flow. In: Shock Compression of Condensed Matter - 2005, Eds .: M.D. Furnish et al. - New York: American Institute of Physics, 2006. - P. 880–883.
66. Kanel G.I., Nellis W.J., Savinykh A.S., Razorenov SV. and Rajendran A.M., J. Appl. Phys. 2009. Vol. 106. P. 043524.
67. Golkov R., Kleiman D. and Zaretsky E.B., Impact response of single crystal potassium chloride at elevated temperatures. In: Shock Compression of Condensed Matter - 2003, Eds. M.D. Furnish et al. AIP Conference Processions. New York: American Institute of Physics. 2004. Vol. 706. P. 735–738.
68. Bundy F.P., Basset W.A., Weathers M.S., Hemley R.J., Mao H.K., and Goncharov A. F., J. Chem. Phys. 1996. Vol. 34 (2). P. 141–153.
69. Turkevich V.Z., High Pressure Research. 2002. Vol. 22 (3). P. 525–529.
70. Kurdyumov A.V., Britun V.F., Borimurchuk N.I., Yarosh V.V., Martensitic and sieve and diffusion transformations in carbon and boron nitride at shock compression. - Kiev: Kupriyanov Publishers, 2005..
71. Bundy F P. and Kasper J.S., J. Chem. Phys. 1967. Vol. 46 (9). P. 3437–3446.
72. Kanel G.I., Bezruchko G.S., Savinykh A.S., Razorenov S.V., Milyavskii V. V., Khischenko K.V., Teplofizika Vysokikh Temperatur. 2010. Vol. 48, No. 6. P. 845–853.
73. Garkushin G.V., Kanel G.I., Razorenov S.V., Savinykh A.S., Izv. RAN. Mekh. Tverd. Tela. 2017. No 4. P. 69–79.
74. McQueen R.G., Fritz J.N., and Morris C.E.. The velocity of sound behind strong shock waves in 2024 Al. In: Shock Compression of Condensed Matter - 1983 (edited by J.R. Asay et al.). - Elsevier Sc. Publ. B.V., 1984. - P. 95–98.
75. Kanel G.I., Dremin A.N., Fiz. Goreniya Vzryva. 1977. V. 12, No. 1. P. 85–92.
76. Kanel G.I., Utkin A.V., Fortov V.E., The Equation of State and Macrokinetics De-

composition of Solid Explosives in Shock and Detonation Waves. Soviet Technology Reviews, Section B: Thermal Physics Reviews. 1992. Vol. 3, part 3. P. 1–86.

77. Kanel G.I., Utkin A.V. and Razorenov S.V., Central European Journal of Energetic Materials. 2009. Vol. 6 (1). P. 15–30.
78. Kanel G.I., Razorenov S.V., Fortov V. E. Shock-Wave Phenomena and the Properties of Condensed Matter. - New York: Springer, 2004. - 320 p.
79. Kanel G.I., Fortov V.E., Razorenov S.V., Usp. Fiz. Nauk, 2007. V. 177, No 8. P. 809–830.
80. Kanel G.I., Zaretsky E.B., et al., et al., International Journal of Plasticity. 2009. Vol. 25 (4).
81. Kanel G.I., Int. J. Fract. 2010. Vol. 163 (1–2). P. 173–191.
82. Kanel G.I., Izv. RAN. Mekh. Tverd. Tela. 2014. No 6. P. 6–18.
83. Kanel G.I., Zaretsky E.B., Razorenov S.V., Ashitkov S.I., Fortov V.E., Usp. Fiz. Nauka. 2017. V. 187, No. 5. P. 525–545.
84. Baumung K., Bluhm H. J., Goel B., Hoppe P., Karow H.U., Rush D., Fortov V. E., Kanel G. I., Razorenov S.V., Utkin A.V., and Vorobjev O. Yu.. Laser and Particle Beams. 1996. Vol. 14, No. 2. P. 181–210.
85. Crowhurst J.C., Reed B.W., Armstrong M.R., Radousky H.B., Carter J.A., Swift, D. C., Zaug, J.M., Minich, R.W., Teslich, N.E., and Kumar, M.,J. Appl. Phys. 2014. Vol. 115. P. 113506.
86. Ashaev V.K., Doronin G.S., Levin A.D., Fizika Goreniya Vzryva. 1988. V. 24, No. 1. P. 95–99.
87. Utkin A.V., Pershin S.V., Fortov V.E., DAN, 2000. V. 374, No. 4. P. 486–488.

Index